THE JOHN HARVARD LIBRARY

Howard Mumford Jones
Editor-in-Chief

A Woman Rice Planter

By
PATIENCE PENNINGTON

Edited by Cornelius O. Cathey

THE BELKNAP PRESS OF
HARVARD UNIVERSITY PRESS
Cambridge, Massachusetts

1961

© Copyright 1961 by the President and Fellows of Harvard College
All rights reserved

Distributed in Great Britain by Oxford University Press, London

Library of Congress Catalog Number 61-13741

Printed in the United States of America

CONTENTS

INTRODUCTION BY CORNELIUS O. CATHEY   xi

A WOMAN RICE PLANTER

CHAPTER I	1	CHAPTER VIII	244
CHAPTER II	59	CHAPTER IX	282
CHAPTER III	94	CHAPTER X	325
CHAPTER IV	128	CHAPTER XI	349
CHAPTER V	178	CHAPTER XII	374
CHAPTER VI	197	CHAPTER XIII	381
CHAPTER VII	221	CHAPTER XIV	406

ILLUSTRATIONS

The sheaves are beaten with flails *facing page*	1
"Cherokee" — my father's place	4
Bonaparte	7
Each field has a small floodgate, called a "trunk"	9
Marcus began work on the breaks	10
"The girls shuffled the rice about with their feet until it was clayed"	11
Near the bridge two negro women are fishing	14
A request from Wishy's mother, Annette, for something to stop bleeding	17
Green thought it was folly and fussiness	27
She picked her usual thirty-five pounds alone	31
Today the hands are "toting" the rice into the flats	34
"You see a stack of rice approaching, and you perceive a pair of legs, or a skirt, as the case may be, peeping from beneath"	35
Pallas	40
Front porch — Casa Bianca	42
Elihu was a splendid boatman	51
My little brown maid Patty is a new acquisition and a great comfort, for she is very bright	53
The roughness and plainness of the pineland house	54
The yearly pow-wow at Casa Bianca	60
"Four young girls who are splendid workers"	62
She promised not to war any more	65
"Myself, ma'am, bin most stupid"	66
A rice field "flowed"	72
The hoe they consider purely a feminine implement	79
The back steps to the pineland house	84
"A very large black hat"	87
Her husband brought her in an ox cart	93
"Old Maum Mary came to bring me a present of sweet potatoes"	98

ILLUSTRATIONS

"Pa dey een 'e baid"	102
One or two hands in the barnyard	107
A corner of Casa Bianca	109
"Chaney"	112
Five children asked me to let them "hunt tetta"	120
"It is tied into sheaves, which the negroes do very skilfully, with a wisp of the rice itself"	122
"The field with its picturesque workers"	124
"The Ferry"	132
His wife was very stirring	136
Day after day I met Judy coming out of her patch	138
"Old Florinda, the plantation nurse"	144
"Miss Patience, le' me len' yer de money"	150
"Jus' shinin' um up wid de knife-brick"	159
Aphrodite spread a quilt and deposited the party upon it	164
"Then he could talk a-plenty"	171
Chloe is devoted to the chicks — feeds them every two hours	174
Prince Frederick's Pee Dee	178
Prince George Winyah	180
"Eh, eh, I yere say yu cry 'bout chicken"	187
The summer kitchen at Cherokee	188
The winter kitchen at Cherokee	189
The string of excited children	190
I got Chloe off to make a visit to her daughter	198
I really do not miss ice, now that my little brown jug is swung in the well	200
Patty came in	210
"Plat eye!"	216
Goliah cried and sobbed	225
Had Eva to sow by hand a little of the inoculated seed	232
Her little log cottage was as clean as possible	236
The sacred spot with its heavy live oak shadows	242
"I met Dab on the road"	249
Cherokee steps	250
The smokehouse at Cherokee for meat curing	260
Sol's wife, Aphrodite, is a specimen of maternal health and vigor	262
I saw a raft of very fine poplar logs being made	263
Cypress trees	265
She was a simple, faithful soul — always diligent	270
Winnowing house for preparation of seed rice	272

ILLUSTRATIONS

"Patty en Dab en me all bin a eat"	276
Chloe began: "W'en I bin a small gal"	288
I took Chloe to Casa Bianca to serve luncheon	299
"I read tell de kumfut kum to me"	309
"Up kum Maum Mary wid de big cake een de wheelbarrer"	311
Gibbie and the oxen	313
In the field — sowing	317
How to lay the breakfast table	321
Joy unspeakable	326
The church in Peaceville	331
Chloe was a great success at the North	338
My old summer home at Pawleys Island	349
The roof of the house on Pawleys Island — from the sandhills	352
"En de 'omans mek answer en say: 'No, ma'am; we neber steal none'"	356
"Dem all stan' outside de fence"	367
Fanning and pounding rice for household use	375
Pounding rice	376
The rice fields looked like a great lake	399
Casa Bianca	422
Rice fields from the highlands	439
"You see I didn't tell no lie"	442

INTRODUCTION

The culture of rice in what is now the United States was first undertaken in Virginia, but the first successful planting of the crop occurred in South Carolina in the latter half of the seventeenth century.[1] Although cultivation of this crop spread soon to North Carolina and Georgia, and later to Florida, Alabama, Mississippi, and Louisiana, South Carolina produced from sixty to eighty per cent of the rice grown in this country down to the end of the Civil War. The emancipation of the slaves and other causes led to a gradual decline of interest in the crop after 1865; and, at the present day, South Carolina has altogether abandoned the production of rice. The book here reproduced recounts the heroic efforts, over a four-year period beginning in 1903, of the daughter of one of the outstanding ante-bellum South Carolina rice planters to carry on the culture of the crop on the ancestral lands, using free labor, most of the laborers being descendants of her father's former slaves. That her efforts ended in failure need not be attributed to a lack of knowledge of planting procedures or want of diligence in management or of perseverance, but rather to economic, geographic, and human factors over which she had no control. A basis for understanding this failure may be found in a brief survey of the history of the rice industry as it was carried on in the ante-bellum South and of developments within the industry since 1865.

After some years of experimentation when the crop was

[1] Lewis Cecil Gray, *History of Agriculture in the Southern United States to 1860*, 2 vols. (Washington: Published by the Carnegie Institution of Washington, 1933), I, 277-279.

planted in the inland, fresh-water swamp lands which had been cleared for that purpose, the center of rice culture shifted to lands nearer the coast, where tidal action would facilitate the necessary flooding of the fields and the crop would be less endangered by freshets. The area around Georgetown, South Carolina, was particularly well adapted by nature for the planting of rice as it was done in the time of slavery. There the Sampit, Black, Peedee, and Waccamaw rivers converge to form Winyah Bay. A few miles to the south, the Santee flows directly into the Atlantic. Along the banks of these rivers and as far up as the effect of the tide was felt, rice became the "grand staple" of plantation production. Here was developed the most productive area of the Rice Coast, the heaviest concentration of slave labor, and a way of life that made the South Carolina low country one of the most interesting and attractive areas of the ante-bellum South.

Unlike the growing of short staple cotton, which could be produced throughout most of the South, rice planting in antebellum times was confined to a narrow strip along the coast. Unlike the growing of cotton too, the culture of rice necessitated such heavy expenditures of labor that its production was necessarily large scale. In the Old South this meant that the plantation — using slave labor with the object of making profits — was the sole producing unit. The success attained by many of these rice planters was reflected in the attractive houses and grounds they maintained for themselves, and by the strong influence they exercised over the economic, political, and social life of the state. They had the money and the inclination to send their sons to the best educational institutions here and abroad, and to provide their daughters with such training as would equip them not only to adorn but also to manage their households. They visited the best watering places, kept up summer houses in the mountains or at the seashore, nor was it unusual for a successful rice planter to take some or all of his family on a prolonged tour of the Continent.

INTRODUCTION xiii

Robert Francis Withers Allston (1801-1864), father of Elizabeth Waties Allston or "Patience Pennington," was one of the most successful of the ante-bellum rice planters. After graduation from West Point in 1821 and a year of duty with the Topographical Service, during which time he did harbor survey work, Robert Allston resigned from the army and returned to South Carolina to assist his widowed mother in the management of her estate. From that time until his death, although rice planting was his primary occupation, he served four years as surveyor-general, twenty-eight years as a member of the legislature, two years as governor (1856-1858) of South Carolina, and gave generously of his time in promoting the economic, social, and cultural life of his community.[2] Through inheritance and purchase he acquired in the Georgetown area seven plantations totaling approximately 4,000 acres, pasture and timber lands in the same area totaling about 9,500 acres, a 1,900 acre plantation in North Carolina, houses in Charleston, Georgetown, Society Hill, Plantersville, and on Pawleys Island, and about six hundred slaves.[3] At the time of his death in 1864, in addition to other bequests he left each of his five surviving children a plantation and approximately one hundred slaves.[4] Unfortunately, the full story of the accumulation of this vast estate cannot be told, for, although Allston was a meticulous record-keeper, most of his papers were wantonly destroyed by his ex-slaves when, in the enjoyment of their new-found freedom at the end of the Civil War, they ransacked the home plantation, Chicora Wood. Patience Pennington, who witnessed the results of that frantic

[2] James Harold Easterby, (ed.), *The South Carolina Rice Plantation as Revealed in the Papers of Robert F. W. Allston* (Chicago: University of Chicago Press, 1945), pp. 11-16.

[3] *Ibid.*, pp. 20-23, 28-29; Elizabeth W. Allston Pringle, *Chronicles of Chicora Wood* (New York: Charles Scribner's Sons, 1922), pp. 209-210.

[4] For an appraisal of the value of this estate, and an account of its disposition, see Easterby, *The South Carolina Rice Plantation*, pp. 43-49.

search for money or treasure wrote: "It was a scene of destruction, and papa's study, where he kept all his accounts and papers, as he had done from the time he began planting as a young man, was almost waist-deep in torn letters and papers." [5]

Robert Allston's experience as a surveyor peculiarly fitted him for the work of reclaiming swamp lands for the planting of rice. Building up the banks to protect the land against flooding in time of freshet or unusually high tides, the location and construction of the trunks to control the flow of water to and from the fields, and the digging of the canals and ditches through which the water passed were but first steps in the reclamation process, steps in which an engineer's skill was put to the test. Once these projects were completed the arduous work followed of clearing the land of trees and undergrowth so that a crop could be planted.[6] In the absence of labor-saving machinery, these undertakings necessitated the expenditure of enormous amounts of labor under difficult and unhealthy circumstances — work it was thought the Negro, conditioned as he was through long generations of life in the hot, steamy jungles of Africa,[7] was better fitted to do than the white man. Once the plantation was established, the problem of maintaining adequately the banks, canals, ditches, and trunks was a continuing one for every rice planter, one that provided the slaves with productive, off-season employment. A break in the water-control system, occasioned by either a freshet or the flooding of the fields with salt water, often resulted in the total loss of the rice crop.

[5] Pringle, *Chronicles of Chicora Wood*, p. 269.
[6] Duncan Clinch Heyward, *Seed From Madagascar* (Chapel Hill: The University of North Carolina Press, 1937), pp. 18–20.
[7] Patrick H. Mell, "Rice Planting in the Agricultural Development of the South," *The South in the Building of the Nation: . . .*, 13 vols. (Richmond: The Southern Historical Publication Society, 1909–1913), V, 175.

INTRODUCTION

Undoubtedly, the Rice Coast was an unhealthy region for both whites and Negroes. The lack of adequate medical services, even for that time, and want of knowledge of how malaria is transmitted resulted in high fatality rates among both races. The unfortunate ones who could not spend their summers either on the seashore or in the highlands generally accepted as a matter of course the attacks of "bilious fever" with which they were afflicted almost every year. In commenting on this situation, Edmund Ruffin, who was the most influential critic of ante-bellum Southern agriculture, wrote: "I cannot but consider the culture of rice as a curse to any region, as it must cause the production of malaria and its consequent diseases. . . ."[8] One planter, near Georgetown, considered the problem of maintaining the health of the work force so important that he sent four of his five sons through medical school and followed this by extra training for them in Paris, not to equip them for the practice of that profession, but to enable them to render better medical service to their slaves.[9] In the usual case, where the plantation was in a remote location, resort was had to home remedies, administered by the planter, his wife, or the overseer, who used medicines from the plantation supply, a doctor being called in only in extreme cases. This was not true at Waverly, one of the Allston plantations. A doctor visited Waverly ninety-two times in the months from January through June and again in November and December 1853, rendering all sorts of medical services, for which his total charge was $390.21. The records do not indicate who rendered such service in the months from July through October.[10] Aside from all other considerations,

[8] *Farmer's Register*, VIII (April 1840), 244.
[9] Pringle, *Chronicles of Chicora Wood*, pp. 157–158.
[10] Easterby, *The South Carolina Rice Plantation*, pp. 342–345. A doctor who regularly visited one of the Allston plantations wrote: "On every well regulated Rice Plantation, the *sick* receive the *first* and if necessary the undivided attention of the overseer." *Ibid.*, p. 349.

the heavy capital investment the planter had in his slaves led him to feel a concern for their health and physical well-being. This is reflected in the instructions the planter gave to his overseers and in the regularity with which the overseers commented in their reports upon the health of "the people." Obvious cases of illness or incapacity to work for any reason posed less of a problem than did cases where malingering was suspected. For example, one of Robert Allston's overseers reported from Waverly: "The People there keeps healthy all but Lidia who I think is playing Posom." Prudence suggested a kind and considerate decision even in this case: "Mrs. Allston has taken her to the seashore." [11]

The general unhealthiness of the Rice Coast in the summer months led many planters to maintain places in the mountains or at the seashore, to which they would repair during the hot weather, leaving the actual supervision of plantation operations at the busiest season of the year to overseers. This practice, of course, was often detrimental to the best interest of the planter and of his laboring force. Robert Allston was more fortunate than most planters in this respect, in that his home plantation, Chicora Wood, was only four miles as the crow flies from his seashore house on Pawleys Island, where the family lived during most of the summer months, and only six miles from Plantersville, where he also had a house for use during the summer. Although the scope of his operations necessitated his using overseers, it was his practice to commute from these places daily to supervise the work being done on the plantation.[12] In a report of 1847 to the South Carolina Agricultural Society, Allston said that the "too general prevalence among planters of absenteeism," which he condemned, was the reason why greater improvements were not being made in rice planting. In the district covered by that report,

[11] *Ibid.*, pp. 254–255.
[12] Pringle, *Chronicles of Chicora Wood*, pp. 72–73.

he found only fifty-one of one hundred and eleven planters resident on their plantations. "Of the major number all were absent, and are habitually absent during the crop season, from some time in the month of May to November frost. Twelve of this number have their residence elsewhere." [13] For the responsibilities they bore, undoubtedly, the overseers of such plantations were the most poorly paid employees in our economic history. Unlike one of his neighbors, who spelled out in finest detail the duties, privileges, and responsibilities of his overseer,[14] Allston used a simpler form of contract, paid his overseers well by the standards of the time, and even provided legacies for two of them in his will.[15] After agreeing "to make Mr. Allston's interest his own," it was the overseer's duty to see to it that the proprietor's plans were carried out.[16] When absent on public business, as Allston frequently was, he transmitted his orders to the overseer through his almost daily correspondence with Mrs. Allston.

To assist the overseer in the management of the labor, a "driver" was selected from among the slaves both to lead and drive the other workers. The "driver" helped turn the workers out in the mornings, shaped up the work gangs, assigned tasks, set the work pace, checked on the completion and quality of work performed, helped issue rations and clothing, "carried

[13] *North Carolina Farmer*, III (February 1848), 193-194. A neighbor and kinsman wrote Allston requesting part-time use of his overseer: "My health is declining, my System waisting, without appetite and Sleep and I know ere long I must terminate my career, unless I can get to the Mountains in time to escape Disease . . . You know My Dear Robert . . . that to remain here a Summer will be Suicide." Easterby, *The South Carolina Rice Plantation*, pp. 250-251.

[14] Ulrich Bonnell Phillips (ed.), *Plantation and Frontier Documents: 1649-1863*, 2 vols. (Cleveland: The A. H. Clark Company, 1909), I, 115-122.

[15] Easterby, *The South Carolina Rice Plantation*, pp. 24-27.

[16] For an interesting sampling of reports and other documents pertaining to the overseers on the Allston plantations, see *ibid.*, pp. 245-330.

the keys" in the absence of the overseer, and, on some plantations, also administered discipline under the supervision of the overseer or proprietor.[17] The "driver" was expected to command the respect of the people who lived on the "street," as the slaves spoke of the area where they lived, and was a person for whom the proprietor felt a special concern. It was a common practice among planters to prohibit the overseer from punishing the "driver" without the owner's prior consent.

Three systems of management were used in directing slave labor. The most popular and perhaps the best for use along the Rice Coast was that in which each worker was assigned a task for the day or week, the satisfactory completion of which gave the slave the remainder of the day or week off. The gang system was one in which the slaves worked together, the "driver" usually setting the pace and insisting that the others keep up. The third, and undoubtedly the least satisfactory, was one in which the worker was simply assigned a task or chore over which there was no supervision or control except that which sprang from a fear of punishment in case the work should be poorly done. All three systems were used on the Allston plantations.

The drainage ditches which divided the rice fields into half- or quarter-acre plots facilitated the assignment of tasks. By the early 1840's standard tasks had been settled upon for practically every kind of work pertaining to the cultivation of rice.[18] To secure the stimulus expected from this system, however, any task was limited in such a way as to offer even the slowest worker the prospect of leisure time as a reward for diligence. As one rice planter put it: "A task is as much work as the meanest full hand can do in nine hours, working industriously . . . This task is never to be increased, and no

[17] Heyward, *Seed From Madagascar*, pp. 157–158.
[18] Ulrich Bonnell Phillips, *American Negro Slavery* (New York: D. Appleton and Company, 1918), pp. 247–248.

INTRODUCTION

work is to be done over task except under the most urgent necessity; which over-work is to be reported to the proprietor, who will pay for it."[19] Robert Allston, like all planters where the task system was employed, classified his slaves, other than the house servants, as "full," "half," and "quarter" hands. He also trained men with special talents or aptitudes for duty as carpenters, cobblers, coopers, blacksmiths, boatmen, or trunk minders, and the women as nurses, cooks, and seamstresses. Prizes were offered each year to stimulate proficiency in these and other types of work, and, undoubtedly, a slave took great pride in being recognized as the prize plowman, or sower, or harvest hand for the year.[20]

Although he was strongly inclined towards religion and active in his church, there is little evidence that Robert Allston questioned the morality of slavery. There is much evidence, however, that he was sincere in believing that the Negroes were better off in servitude than they would be if they were free, and that the white master was obligated to treat them judiciously and well.[21] "Nothing sooner attracts the confidence of the negro," he wrote, "and commands his respect, than the illustration, in a system of management, of justice, tempered by kindness."[22] Writing with detachment of the slaves as "descendants of the African bondsmen given to our ancestors by the mother country," he said: "They are well fed and clothed, well sheltered, and cared for in sickness, and during the infirmities and helplessness of old age. They are for the most part healthy and cheerful, and when well trained, are very efficient laborers."[23] In his judgment, the slaves had

---

[19] *Ibid.*, p. 267.
[20] Pringle, *Chronicles of Chicora Wood*, pp. 12–15; Easterby, *The South Carolina Rice Plantation*, pp. 20, 34.
[21] Easterby, *The South Carolina Rice Plantation*, pp. 16–17.
[22] *DeBow's Review*, XVI (June 1854), 615.
[23] *Ibid.*

provided for them "all the necessaries of life in sufficient abundance. And they enjoy the privilege of procuring many comforts and indulgences." Robert Allston was not the type of man to whom the slaves would feel warmly attached, but he had their respect. It is too much to say that his slaves were happy or contented with their lot. Such comment as the following by one of his overseers would belie the assertion: "I can see since Stephn left a goodeal of obstanetry in Some of the People. Mostly mongst the Woman a goodeal of Quarling and disputeing & teling lies." [24] Although such situations were occasionally the source of trouble, Allston's slaves nevertheless unquestionably felt a sense of security they lost when freedom came. Where slavery had fashioned a bond of obligation between master and slave out of which grew a sense of mutual trust and dependence, freedom snapped that bond and imposed upon the ex-slave the necessity for being self-dependent — a status for which he was very poorly prepared.

Robert Allston came to be regarded as one of the most successful and articulate of the ante-bellum rice planters. His plantations proved his superior talent for organization, and his opinions on rice planting and on the agriculture of his region were widely disseminated in his published articles entitled: "Memoir of the Introduction and Planting of Rice in South Carolina," [25] and "Essay on Sea Coast Crops." [26] At the time of the publication of these essays, an agricultural revolution was running rampant in the American countryside. Simply, the objectives of that revolution were the expansion and improvement of agriculture. Its promotion was through the agricultural press, fairs, and societies which were founded all over the country. Everywhere agriculturists were urged to question the efficiency of their farming practices, to experi-

[24] Easterby, *The South Carolina Rice Plantation*, p. 291.
[25] *DeBow's Review*, I (April 1846), 320–357.
[26] *Ibid.*, XVI (June 1854), 589–615.

INTRODUCTION

ment with new methods, crops, fertilizers, machinery, and to work for the reform of their profession.

Robert Allston was one of the leaders in that movement. He was president of the Winyah and All Saints Agricultural Society for years, an active participant in the work of the South Carolina Agricultural Society, and a frequent contributor to the agricultural press. His emphasis in these capacities was on the progress of science in agriculture, but as a public official, a member of the South Carolina Historical Society, and the Carolina Art Association he also promoted the advance and diffusion of knowledge generally. James D. B. DeBow said that publications like Allston's paper on rice planting, in which he dealt with the history of the crop, the process of reclaiming lands for its use, methods in culture, value as a food and as a commodity in trade, had done more "for the advancement of agricultural science than can well be conceived." [27]

Allston's rice won prizes for its high quality at several expositions including that held in Paris in 1855.[28] In his "Essay on Sea Coast Crops," which was read before the Agricultural Association of the Planting States in 1853, he surveyed the potential of his section for agricultural self-sufficiency, and emphasized the production of crops required by home needs on every plantation — a project dear to the hearts of Southern agricultural reformers and one that Allston succeeded in accomplishing on his own plantations. He was what the less progressive agriculturists contemptuously called a "book farmer."

Robert Allston and his family lived in the style expected of a successful rice planter. Although less inclined than some of his neighbors to spend much time in the social whirl, he belonged to a neighborhood club, "The Hot and Hot Fish Club," the St. Cecilia Society, the South Carolina Jockey

[27] *Ibid.*, I (April 1846), 356.
[28] Easterby, *The South Carolina Rice Plantation*, pp. 92, 123, 149.

xxii INTRODUCTION

Club, the South Carolina Historical Society, and the Order of Masons. The home plantation, Chicora Wood, was beautifully designed for entertaining the family guests. The lands round about abounded in wild life — particularly deer, duck, and wild turkey — which Robert Allston found great pleasure in hunting. Devoted to the cause of religion, he and his family gave generously of their time and resources to promote the work of their church, the Protestant Episcopal, and he made provision for the regular instruction in religious matters, on a voluntary basis, of the Allston slaves.[29] Despite the calls made on his time by these interests, by his civic responsibilities, and by the pressure of plantation affairs, he found time to be a devoted husband and father.

Although Robert Allston left a plantation and approximately one hundred slaves to each of his five surviving children at the time of his death in 1864, circumstances prevented the children from receiving any part of their inheritance. The outcome of the war terminated all property rights in slaves. A few years prior to that event, somewhat contrary to his better judgment, Robert Allston had become heavily indebted to his wife's widowed sister-in-law through buying from her on credit a plantation and one hundred and sixteen slaves. Only the interest had been paid on this indebtedness prior to his death; and since the principal was covered by a mortgage on his landed property, the end of the war found his estate insolvent. Despite heroic efforts by his executors and members of the family to salvage some part of the estate, it was finally disposed of at public auction in 1868. Mrs. Allston was permitted by the creditors to retain Chicora Wood and some adjacent pine lands under her right of dower.[30]

When the war ended Elizabeth Waties Allston was twenty

[29] *Ibid.*, pp. 16, 34.
[30] *Ibid.*, pp. 22, 45, 49; Pringle, *Chronicles of Chicora Wood*, pp. 9–11, 209–211.

years of age. As befitted a young woman of her position in ante-bellum society, Patience, as I shall call her hereafter, had been tutored by an English governess at Chicora Wood until she was nine, when she joined her older sister in attending Madame Togno's boarding school in Charleston. There in company with other girls of about her age and background, Patience studied arithmetic, diction, English, French, history, music, and singing, the instruction being given in French with emphasis upon the fine arts. Except for the summers, when she joined the family on Pawleys Island, she spent the years until nearly the end of the Civil War in that school. Her particular aptitude for languages, literature, and music was stimulated by frequent attendance at concerts or the theatre. This was part of the "finishing off" process through which the young ladies passed — a process in which Madame Togno took great pride.[31]

When the Federal bombardment of Charleston began in July 1863, Patience accompanied her mother to Society Hill, an inland location where the family owned property and where it was thought they would be safe. "Our household consisted only of Mamma," she wrote, "my little sister, and myself, for papa remained at his work on the plantation, only coming now and then for a few days . . . We had the full force of servants . . . Mamma at once began to plant the farm and garden, with the house-servants, and made wonderful crops."[32] Although food continued to be available, other supplies grew increasingly scarce as the war came closer home, and Patience and her mother, like most of the women of the South, learned to use substitutes, to improvise, and to do without. "At Society Hill," she wrote, ". . . the loom was set up in the wash-kitchen, and I learned to weave as well

---

[31] Pringle, *Chronicles of Chicora Wood*, pp. 125–136, 176–183.
[32] *Ibid.*, pp. 189–194.

as to spin, and we knit, knit, knit all the time." [33] It was from this refuge that the family hastened to Chicora Wood when news came that Robert Allston was seriously ill. His death, which occurred a few days after their arrival, left Patience with a "sense of terrible desolation and sorrow," for she was particularly devoted to her father, and regarded him as "the only person in the world in whom I had absolute faith and confidence. I had never seen him show a trace of weakness or indecision. I had never seen him unjust or hasty in his judgment of a person. . . . Never a sign of self-indulgence, or indolence, or selfishness." [34] It was in keeping with these sentiments that she later dedicated *A Woman Rice Planter* to his memory.

The end of the war brought revolutionary change into the lives of the people along the Rice Coast, whites as well as Negroes. In one of her efforts to save at least part of her husband's estate, Mrs. Allston converted her house in Charleston into a boarding and day school where Patience was employed, somewhat against her will, as a teacher of French, literature, and piano. The school was a success, and to her surprise Patience found that she not only had a talent for it but that she enjoyed teaching.[35] Mrs. Allston, however, preferred the country, and in 1869 moved back to Chicora Wood, to which she had obtained full title through exercise of her dower rights, as I have said. By that time social life had become gay again in Charleston, and Patience, understandably, had less enthusiasm for this move than any other member of the family.[36] The plan was that her younger brother, Charles Petigru, who had just graduated from college, be established where he "can make a living by planting

---

[33] *Ibid.*, pp. 195–196.
[34] *Ibid.*, pp. 200–211.
[35] *Ibid.*, pp. 289–291, 316–330.
[36] *Ibid.*, p. 340.

rice." This he succeeded in doing, using free labor. An older brother, Benjamin, who had graduated from West Point before the war and who made a brilliant record in the Confederate Army, also resumed planting rice, but gave it up because of "difficulties with which he was unable to cope." [37] Benjamin later entered the Episcopal ministry.

After the shouting and tumult with which most of the slaves had welcomed freedom had passed, the Negro soon found it necessary again to work for the white man. A few felt compelled to exercise their new-found right to move, and for sometime thereafter lived as vagabonds searching through the land for greener pastures. Most, however, remained in the same communities in which they had lived "befo' de wah," and became sharecroppers, tenants, and wage hands on the old plantations. This was the case with most of the Allston slaves, who provided the labor supply for Benjamin and Charles Petigru Allston's rice planting operations after the war.

Unfortunately for the slave, freedom was not undergirded by any grant of property, or by a plan designed to encourage his progress toward economic self-sufficiency. He had no land, no property, and very little experience that would permit him to launch out on his own with a reasonable chance for success. His chief asset was his ability to labor. On the other hand, his former master was in a somewhat similar plight. His plantation was a shambles, his livestock had been either driven away or slaughtered, his tools were either worn out or missing, his capital was exhausted, he had very little credit, and he now had no assured labor supply. It would seem that these hard facts alone would be sufficient to bring the white man and the Negro together again, this time in a landlord-and-tenant relationship. There was, however, an-

[37] Easterby, *The South Carolina Rice Plantation*, p. 18.

other compelling force. The bond of mutual dependence which bound one to the other in the slave regime was not entirely an economic one, it was also a humanitarian one which freedom did not sever and which continued in the genuine concern felt by each for the other's well-being. While on a visit to the plantation in North Carolina, just after Sherman's army had passed through, one of the elderly slaves said to Mrs. Allston that he wanted to go "home." To her inquiry as where his home was, he replied: "Wherever you is, miss, dere is my home." [38] "I keep track of all the descendants of our family servants," Patience Pennington wrote many years later, "and it gives me a great pleasure when they make good and do credit to their ancestry." [39]

Efforts were made all along the Rice Coast to resume the large-scale planting of rice under the changed circumstances. "The fust contract was," an ex-slave stated, "you fu'nish land and seed and animals an' get two-thirds; I fu'nish wuck and get one-third. Every day I didn't wuck was deduct' from my share." [40] This arrangement contemplated, of course, the Negro being given a place to live, and rights to firewood, hunting, and fishing. Later, when the Negro became more prosperous and provided his own work animals, the crop was divided in half. Some chose to work entirely for wages, which for all types of employment, but particularly in agriculture, were very low in the post-bellum South.

Such were the working arrangements in the rice industry when Patience Pennington, like her mother and grandmother before her, undertook the role of a woman rice planter. Possessed of a "sensitive temperament and fierce Huguenot conscience," she looked back upon slavery as "a terrible misfortune," and expressed herself as truly thankful that the in-

---

[38] Pringle, *Chronicles of Chicora Wood*, p. 364.
[39] *Ibid.*, p. 61.
[40] *Ibid.*, p. 366.

stitution was a thing of the past.[41] Lacking the control of labor that slavery gave the master, and with pitifully inadequate financial resources, she could only appeal to the laborers' self-interest, and, as she expressed it, "give them a chance to work for themselves and prove themselves worthy to be free men." The fact that she later abandoned planting does not prove that the Negroes were unworthy to be free, but is attributable to a combination of circumstances and developments with which Patience was unable to contend. Under slavery, there had always been someone to do the Negro's planning for him, so that his efforts were directed by a "compelling hand." Generally the slaves accepted this circumstance and tended to associate their own well-being with that of their masters. Their sense of security was enhanced as they demonstrated a willingness and a capacity to work and to bear responsibilities. The thin veneer of self-discipline that the Negro attained in slavery, however, left him poorly prepared for freedom, and many experienced a decline in willingness to accept responsibilities and to work. As master of his own destiny, but weighted down by many handicaps, the freedman was not usually stimulated to become a very respectable member of the society in which he lived. Unfortunately for him, nature had prescribed that the living be easy along the Rice Coast.

Left a widow six years after her marriage to John Julius Pringle, Patience entered upon her rice planting venture by purchasing, partially on credit, White House plantation on the Pee Dee River. This plantation, the gardens of which had been planted by the famous botanist-diplomat, Joel R. Poinsett, had belonged to her husband's family, and it was there that she had spent her short married life.[42] Although herself skeptical of a woman's ability to handle such an un-

[41] *Ibid.*, pp. 7–8.
[42] Easterby, *The South Carolina Rice Plantation*, p. 18.

dertaking, she found the Negroes, most of whom were formerly Allston slaves or their descendants, willing to work under the arrangements she prescribed. Some were employed as wage hands, and others as sharecroppers paying ten bushels of rice per acre for the lands allotted to them. There was no overseer. The methods employed in the cultivation of the crop were the same as those used before the war. In the beginning, both parties profited from their endeavors; Patience paid off the mortgage on the plantation, and the Negroes made enough to buy such luxuries as a yoke of oxen or a horse and buggy.

When her mother died in 1896, Chicora Wood plantation, which is eight miles from White House, had to be sold in order to settle the estate. Patience, whether actuated by a strong sentimental attachment to the place or her father's yen for expansion, bought it — again on credit — rather than see it go to someone outside the family. This decision, as she later admitted, "does more credit to my heart than head, and it is very doubtful if I shall ever pay off the mortgage." Henceforth her planting operations on both plantations were carried out from Chicora Wood where she resided for years, isolated from white neighbors but possessed of a capacity, although alone, for living a full and interesting life.

Robert Allston once characterized the planter as a conservative by nature, and less inclined to act in combination with other planters than any other description of men doing the same kind of work. He wrote:

Bred in the country amid the exuberance of nature, in her just proportions and distribution, his mind is accustomed to her gradual processes, the regular succession of the seasons and the annual recurrence of the routine of labor allotted by the Creator. His gifts acquire strength, character, and virtue . . . In the creation around him, even of inanimate nature,
"He finds tongues in trees, books in the running brooks,
  Sermons in stones, and good in everything." [43]

[43] *DeBow's Review*, XVI (June 1854), 591.

INTRODUCTION

Patience Pennington possessed these qualities in good measure, along with a love of music, literature, family, tradition, and a compassion for her fellow man, all of these sources of strength in time of need. Her philosophy is best illustrated by her statement: "If I were asked for what I was most thankful in my possessions I should say my power of enjoyment." She expressed amazement, while on one of her frequent trips to the North, that the simple life was becoming impossible for most people, subjected as they were to the tyranny of fashion and to the rush and hurry of modern living. Even today Chicora Wood, which has long since ceased to be a rice-producing plantation, abounds in those "pretty sights and sounds which nature so lavishly provides," and which gave Patience a feeling of exhilaration. The beautiful setting and the well-kept buildings and grounds combine to create there an air of serenity.

Patience acquired the "diary habit" as a young girl when she was required by her mother, as an exercise in writing, to keep a journal of all that had happened the previous day. The continuance of this practice in her later years not only provided an outlet for the expression of her feelings and experiences as a rice planter, but also the source from which this book sprang. Her diary notations recount with fine sentiment the story not only of her trials, vexations, and triumphs, but also of the death of an old and significant industry, and the end of an era in our economic and social history. This story, told in "a manner of charming simplicity and without a trace of self-consciousness or self-assertion," was first published by the New York *Sun* between 1904 and 1907, and was later compiled in book form as *A Woman Rice Planter,* illustrated by many well drawn vignettes of persons and places mentioned in the account. Selecting for herself the pen name, Patience Pennington, she also sought to hide the identity of some of the sites and people mentioned in the book. For examples, Georgetown becomes "Gregory," Chicora Wood becomes "Cherokee," White House plantation is changed to "Casa Bianca," and Charleston becomes "Carrollton." The author, who is affectionately remembered today by the elder residents of the Georgetown area, also wrote *Chronicles of Chicora Wood*, which was published under her real name in 1922 and which is a general account of the "beauty and pathos" of life in the Allston family until the end of the Civil War.

The great storm of 1906, which destroyed the banks and floodgates along the Rice Coast, was followed by equally devastating weather in the two years following. These three bad crop-years sounded the death knell of rice production along the South Atlantic coast. It was entirely impractical for the banks to be restored by hand labor, as they had been constructed in the slave era, and few, if any, planters had the capital to purchase machinery that would do the work.[44] In the absence of repairs the fields on which the crop had been grown either remained flooded and became a natural feeding ground for all sorts of wild fowl, or gradually became marsh land. Now most of the old plantations have found new owners who maintain them not for agricultural production but as winter homes, as hunting preserves, or as places of refuge from the strain and hurry of modern living.[45] Robert Allston and Patience Pennington would be surprised to know that today, as far as income is concerned, the most valuable lands along the old Rice Coast are those that produce pulpwood — the pinelands, which in their day had had little value.

Prior to 1906, when Patience Pennington abandoned her role as a planter, the nation had begun to turn for its supply

[44] Pringle, *Chronicles of Chicora Wood*, pp. 16–17. See also Heyward, *Seed From Madagascar*, pp. 221–248.

[45] For an excellent account of the rice plantations in the Georgetown area that are still in existence, see Alberta Morel Lachicotte's beautifully illustrated volume, *Georgetown Rice Plantations* (Columbia: The State Commercial Printing Company, 1955).

of rice to the coastal prairie area of Louisiana, Texas, and Arkansas, where lands suitable for the production of the crop exist in abundance. There the fields were far less subject to untimely flooding by freshet or storm, the necessary water could be easily and cheaply provided, and better still, the soil was adapted to machine cultivation. There, one laborer using modern machinery could tend ten times as many acres as one laborer was expected to tend along the old Rice Coast. The efficient combination of capital, labor, and management worked out in the production of the crop in this new area enabled the producer to sell his surplus rice on the world market, even in the Orient, at a profit.[46] Patience Pennington would be pleased to note that much of it goes to market bearing the brand name, "Carolina Rice."

Although the era with which it is immediately concerned has long since passed, the reader will find *A Woman Rice Planter* refreshing as a piece of literature for it is more than a commentary on rice planting — it is essentially a study of man in his relations with nature, with other human kind, and of the human spirit. These are the realities that live on, for Patience Pennington in her narrative succeeds in presenting accounts of even commonplace events, developments, and persons with vividness equal to that of the good artist who embodies his impressions on canvas. The word pictures she paints effectively reflect her innate love of beauty and order, her distaste for ugliness and disorder. Whether consciously done, or not, nearly every facet of her character is revealed by the tenderness, sympathy, and modesty with which she treats her subjects.

Thinking of herself as "the philosopher who, battered and

---

[46] Rupert B. Vance, *Human Geography of the South: A Study in Regional Resources and Human Adequacy* (Chapel Hill: The University of North Carolina Press, 1932), pp. 214–219; Heyward, *Seed From Madagascar*, pp. 212–220.

bruised by life's battles, looks with calm, serene eyes on the stormy path behind her," Patience Pennington sought in her writings to interest her parents' grandchildren (she had no children of her own) in what she called "the pageant of the Past." Possessed of a sensitive and sympathetic nature, she wrote of the past with feeling. With her the past was alive — something to be reflected upon and preserved. In her memory, the past was filled with "scenes as intense as the flaming sky, incidents as tender as the fleecy clouds, years as dark and tragic as that leaden storm-bank at the horizon's edge." [47] Hers is a story of courage, sacrifice, devotion to the land and its people, of success, and of failure — told without complaint or bitterness, but with an attractive simplicity of style.

CORNELIUS O. CATHEY

Chapel Hill, North Carolina

[47] Pringle, *Chronicles of Chicora Wood*, p. v.

A NOTE ON THE TEXT

The text of the John Harvard Library edition of *A Woman Rice Planter* has been reprinted from the original plates, which have been preserved through the years by the Carolina Art Association of Charleston, South Carolina. Consequently the few typographical errors of the original edition reappear here, but in no case is the text rendered illegible by these errors, which are all a matter of single missing or broken letters.

The following glossary of some frequently used Gullah words first appeared at the end of the original edition, and was undoubtedly compiled by Patience Pennington.

unna	you
een	in
ne	in the
fremale	female
tissic	asthma
tetta	potatoes
fai'	fairly
bittle	food, victuals
castle	coffin, casket

The sheaves are beaten with flails.

A Woman Rice Planter

By
PATIENCE PENNINGTON

ILLUSTRATIONS BY
ALICE R. H. SMITH

To
MY FATHER

TO WHOSE EXAMPLE OF SELF-CONTROL AND CHRISTIAN
FORTITUDE, I OWE THE POWER TO LIVE MY
LIFE INDEPENDENT OF EXTERNALS, I
DEDICATE THESE FRAGMENTARY
RECORDS, ON THIS THE ONE
HUNDRED AND TWELFTH
ANNIVERSARY OF
HIS BIRTH

CHICORA WOOD,
    April 21st, 1913.

A WOMAN RICE PLANTER

CHAPTER I

CHEROKEE, March 30, 1903.

YOU have asked me to tell of my rice-planting experience. and I will do my best, though I hardly know where to begin.

Some years ago the plantation where I had spent my very short married life, Casa Bianca, was for sale, and against the judgment of the men of my family I decided to put $10,000, every cent I had, in the purchase of it, to grow old in, I said, feeling it a refuge from the loneliness which crushed me. Though opposed to the step, one of my brothers undertook very kindly to manage it until paid for, then to turn it over to me. I had paid $5000 cash and spent $5000 in buying mules, supplies, ploughs, harrows, seed rice, etc., necessary to start and run the place. This left me with a debt of $5000, for which I gave a mortgage. After some years the debt was reduced to $3000, when I awoke to the fact that I had no right to burden and worry my brother any longer with this troublesome addition to his own large planting,[1] and I told him the first of January of 18— that I had determined to relieve him and try it myself. He seemed much shocked and surprised and said it was impossible; how was it possible for me, with absolutely no knowledge of planting or experience, to do anything? It would be much wiser to rent. I said I would gladly do so, but who would rent it? He said he would give me $300 a year for it, just to assist me in this trouble, and I answered that that would just pay the

[1] He planted at this time one thousand acres of rice successfully.

taxes and the interest on the debt, and I would never have any prospect of paying off the mortgage, and, when I died, instead of leaving something to my nieces and nephews, I would leave only a debt. No; I had thought of it well; I would sell the five mules and put that money in bank, and as far as that went I would plant on wages, and the rest of the land I would rent to the negroes at ten bushels to the acre. He was perfectly dismayed; said I would have to advance heavily to them, and nothing but ruin awaited me in such an undertaking.

However, I assembled the hands and told them that all who could not support themselves for a year would have to leave the place. With one accord they declared they could do it; but I explained to them that I was going to take charge myself, that I was a woman, with no resources of money behind me, and, having only the land, I intended to rent to them for ten bushels of rice to the acre. I could advance nothing but the seed. I could give them a chance to work for themselves and prove themselves worthy to be free men. I intended to have no overseer; each man would be entirely responsible for the land he rented. "You know very well," I said, "that this land will bring my ten bushels rent if you just throw the seed in and leave it, so that every stroke of work that you do will go into your own pockets, and I hope you will prove men enough to work for that purpose."

Then I picked out the lazy, shiftless hands and told them they must leave, as I knew they would not work for themselves. All the planters around were eager for hands and worked entirely on wages, and I would only plant fifty acres on wages, which would not be enough to supply all with work. My old foreman, Washington, was most uneasy and miserable, and questioned me constantly as to the wisdom of what I was doing. At last I said to him: "Washington, you do not know whether I have the sense to succeed in this thing,

Mass' Tom does not know, I don't know; but we shall know by this time next year, and in the meantime you must just trust me and do the best you can for me."

It proved a great success! I went through the burning suns all that summer, twice a week, five miles in a buggy and six in a boat! I, who had always been timorous, drove myself the five miles entirely alone, hired a strange negro and his boat and was rowed by him to Casa Bianca plantation. Then, with dear old Washington behind me, telling of all the trials and tribulations he had had in getting the work done, I walked around the 200 acres of rice in all stages of beauty and awfulness of smell.

But I was more than repaid. I paid off the debt on the place and lifted the mortgage. I had never hoped for that in one year. My renters also were jubilant; they made handsomely and bought horses and buggies and oxen for the coming year's work. When I had paid off everything, I had not a cent left in the bank to run on, however. Washington was amazed and very happy at the results, but when I said something to him about preparing the wages field for the coming crop, he said very solemnly: "Miss, ef yo' weak, en you wrastle wid a strong man, en de Lo'd gie you strenf fo' trow um down once, don't you try um 'gain." I laughed, but, remembering that I would have to borrow money to plant the field this year, I determined to take the old man's advice and not attempt it. This was most fortunate, for there was a terrible storm that autumn and I would have been ruined. My renters were most fortunate in getting their rice in before the storm, so that they did well again.

From that time I have continued to plant from 20 to 30 acres on wages and to rent from 100 to 150 acres. Of course I have had my ups and downs and many anxious moments. Sometimes I have been so unfortunate as to take as renters those who were unfit to stand alone, and then I have suffered

"Cherokee" — my father's place.

serious loss; but, on the whole, I have been able to keep my head above water, and now and then have a little money to invest. In short, I have done better than most of my neighbors.

Five years ago the head of our family passed away, and the Cherokee plantation, which my father had inherited from his grandfather, had to be sold for a division of the estate. None of my family was able to buy it, and a syndicate seemed the only likely purchaser, and they wanted to get it for very little. So I determined the best thing I could do was to buy it in myself and devote the rest of my life to keeping it in the family, and perhaps at my death some of the younger generation would be able to take it. This would condemn me to a very isolated existence, with much hard work and anxiety; but, after all, work is the greatest blessing, as I have found. I have lived at Cherokee alone ever since, two miles from any white person! With my horses, my dogs, my books, and piano, my life has been a very full one. There are always sick people to be tended and old people to be helped, and I have excellent servants.

My renters here, nearly all own their farms and live on them, coming to their work every day in their ox-wagons or their buggies; for the first thing a negro does when he makes a good crop is to buy a pair of oxen, which he can do for $30, and the next good crop he buys a horse and buggy.

The purchase of Cherokee does more credit to my heart than head, and it is very doubtful if I shall ever pay off the mortgage. I have lost two entire crops by freshet, and the land is now under water for the third time this winter, and, though I have rented 125 acres, it is very uncertain if I can get the half of that in. March is the month when all the rice-field ploughing should be done. The earliest rice is planted generally at the end of March, then through April, and one week in May. Last season I only got in fifty acres of rent rice and ten of wages; for in the same way the freshet was over the rice land all winter, and when it went off, there was only time to prepare that much. The renters made very fine crops — 30, 40, and 45 bushels to the acre, while the wages fields only made 17! This is a complete reversal of the ordinary results, for I have very rarely, in all these years, made less than 30 bushels to the acre on my fields, and I was greatly discouraged and anxious to understand the reason of this sudden failure in the wages rice at both plantations.

By the merest chance I found out the cause. Early in December I was planting oats in a six-acre field. We broadcast winter oats in this section and then plough it in on fields which have been planted in peas before. I was anxious to get the field finished before a freeze, and had six of the best ploughmen in it. Grip had prevented my going out until they had nearly finished, but Bonaparte had assured me it was being well done. When I went into the field, it looked strange to me — the rich brown earth did not lie in billowy ridges as a ploughed field generally does. Here and there a weed skeleton stood erect. I tried to pull up one or two of these and found they were firmly rooted in the soil and had never been turned. I walked over that field with my alpenstock for hours, and found that systematically the ploughmen had left from eight to ten inches of hard land between each furrow, covering it skilfully with fresh earth, so that each hand who had been paid for an acre's ploughing had in reality ploughed only one-third of an acre. And then I understood the failure of all the wage rice!

I called Bonaparte, my head man, whom I trust fully. His grandfather belonged to my grandfather, and his family hold themselves as the colored aristocracy of this country. He has been a first-class carpenter, but he is rheumatic and does not work with ease at his trade now, and prefers taking charge of my planting as head man, or agent, as they now prefer to call it. He is trustworthy and has charge of the keys to my barns where rice, corn, oats, and potatoes are kept. I have trusted him entirely, and it would be a dreadful blow to think that he was losing his integrity. Though the pressure from the idle, shambling, trifling element of his race is very great, he has been able to resist it in the past.

I showed Bonaparte what I had discovered, and he seemed terribly shocked. Whether this was real or not I cannot say, but it seemed very real, and as he has never ploughed, perhaps he really did not understand. When I said: "And this is why the wage rice turned out so badly! You received ploughing like this and I paid for it," he seemed convicted and humbled. He had told me how beautifully the rice got up, but as soon as the hot suns of July struck it, the leaves just wilted. Of course, the roots could not penetrate the packed, unbroken clay soil. The best rice-field soil is a blue clay which the sun bakes like a brick. For a while the roots lived in the fresh earth on top.

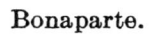
Bonaparte.

The seed rice I had paid $1.35 a bushel for and planted two and one-half bushels to each acre; the cost of cultivating and harvesting it is $15 the acre, so that makes $18.37 which it cost to produce seventeen bushels of rice, which sold at 80 cents a bushel, $13.60.

What is to be the result of this new departure in the way of dishonesty I do not know. It has taken me a long time to lose patience. A few years ago one could get the value of the money paid for work. Just after the war there was a splendid body of workers on this plantation, and every one in the neighborhood was eager to get some of the hands from here. My father gave prizes for the best workers in the different processes, and they felt a great pride in being the prize ploughman or ditcher or hoe hand of the year; but now, alas, poor things, they have been so confused and muddled by the mistaken ideas and standards held out to them that they have no pride in honest work, no pride in anything but to wear fine clothes and get ahead of the man who employs them to do a job.

It is very hard for me to say this; I have labored so among them to try to elevate their ideals, to make them bring up their children to be honest and diligent, to make them still feel that honest, good work is something to be proud of. Even last year I would not have said this, but, alas, I have to say it now.

I have just come in from the corn-field, where two women have been paid for cutting down the corn-stalks, so that there will be nothing to interfere with the plough. They have only broken off the tops of the stalks, leaving about eighteen inches of stout corn-stalks all through the field. I shall have to send some one else to do the work and pay once more.

Yesterday I drove eight miles to my lower place, Casa Bianca, where the foreman asked me to go round the banks with him and see the inroads of the last full-moon tides, and it was appalling, the forces of nature are so immense. It makes me quail to think of the necessity of setting my small human powers in opposition. The rice-field banks are about three feet above the level of the river at high water, and each field has a very small flood-gate (called a trunk), which opens and closes to let the water in and out; but when a gale or freshet comes, all the trunk doors have to be raised so as not to strain the banks, and the water in the fields rises to the level of the river outside.

I must stop writing now or I will get too blue. I must go out and bathe in the generous sunshine and feast my eyes on the glory of yellow jessamine that crowns every bush and tree and revel in the delicious perfume as my bicycle glides over the soft, brown pine-needles along the level paths where the great dark blue eyes of the wild violets look lovingly up at me.

Yes, yes, God is very good and His world is very beautiful, and we must trust Him. When these brown children of His were wild, they were, no doubt, in a physical way perfect, but when they were brought to a knowledge of good and evil and brought under the law, like our first parents, the Prince of Darkness stepped in and the struggle within them of the forces of heaven and hell has been going on there ever since.

Each field has a small flood-gate, called a "trunk."

Can we doubt which will conquer in the end? No! Evil can never have the final victory, but the struggle will be long, for the Prince of Darkness uses such subtle emissaries. They come in the guise of angels, as elevators and instructors, taking from them the simple first principles of right and wrong which they had grasped, and substituting the glamour of ambition, the desire to fly, to soar, for the God-given injunction, "What doth the Lord require of thee, but to do justly and to love mercy and to walk humbly with thy God?" Thank God, there is one man of their own race striving to hold up true standards of the Cross instead of the golden calf of the politician.

I fear this is a dull letter, but I have tried to make you understand something of the situation.

PATIENCE PENNINGTON.

CHEROKEE, June 1, 1903.

Since last I wrote I have been the sport of winds and waves. This place is still under water from a freshet, and on Sunday, April 5, there was a severe gale, and the water swept over the whole 200 acres of Casa Bianca, flowing up the rice-fields in an hour. Saturday evening the hands, after ploughing, left their ploughs in the field to continue work Monday, and they could not see the handles of the ploughs Sunday morning. I went down Tuesday,

Marcus began work on the breaks.

to find bridges carried away and even the banks still under water, and the head man reported five breaks in the Black River bank. It was impossible to do anything until the tide receded, and as there was a strong east wind blowing and a freshet coming down the Pee Dee, things looked very black.

I could not help lamenting aloud, and Marcus felt obliged to offer me some comfort, so he said: "Miss, if we one been

"The girls shuffled the rice about with their feet until it was clayed."

a suffer, I'd feel bad, but eberybody bank bruk, en eberybody fiel' flow." This did not comfort me at all, but I realized the folly of lamenting. Fortunately I had just bought 3000 feet of boards, and as soon as the water left the fields Marcus began work on the breaks, and by driving puncheons, laying ground logs, and flatting mud and filling in, the bank is up again, keeping out the river, and the fields are drying off. The season, however, has not waited on us. April is gone, and not an acre is planted when I expected to have 100 acres growing by this time. The worst is that I have been paying out heavily every week to put things back where they were at the end of March.

There are many curious things about the planting of rice. One can plant from the 15th of March to the 15th of April, then again from the 1st to the 10th of May, and last for ten days in June. Rice planted between these seasons falls a prey to birds, — May-birds in the spring and rice-birds in August and September. It was impossible to plant in April this year, and now every one is pushing desperately to get what they can in May.

Yesterday I went down to give out the seed rice to be clayed for planting to-day. I keep the key to the seed-rice loft, though Marcus has all the others. I took one hand up into the upper barn while Marcus stayed below, having two barrels half filled with clay and then filled with water and well stirred until it is about the consistency of molasses. In the loft my man measured out thirty-five bushels of rice, turning the tub into a spout leading to the barn below, where young men brought the clay water in piggins from the barrel and poured it over the rice, while young girls, with bare feet and skirts well tied up, danced and shuffled the rice about with their feet until the whole mass was thoroughly clayed, singing, joking, and displaying their graceful activity to the best advantage. It is a pretty sight. When it is completely covered with clay, the rice is shovelled into a pyramid and left to soak until the next morning, when it is measured out into sacks, one and one-fourth bushels to each half acre. Two pairs of the stoutest oxen on the plantation are harnessed to the rice-drills, and they lumber along slowly but surely, and by twelve o'clock the field of fourteen acres is nearly planted.

It is literally casting one's bread on the waters, for as soon as the seed is in the ground the trunk door is lifted and the water creeps slowly up and up until it is about three inches deep on the land. That is why the claying is necessary; it makes the grain adhere to the earth, otherwise it would float. Sometimes, generally from prolonged west winds, the river is low, and water enough to cover the rice cannot be brought in on one tide, and then the blackbirds just settle on the field, diminishing the yield by half.

I went down into the Marsh field, where five ploughs are running, preparing for the June planting. It is a 26-acre field, very level and pretty, and I am delighted with the work; it is beautiful. When I told one of the hands how pleased I was with the work, he said: "Miss, de lan' plough so sweet, we haf for do' um good." I went all through with much pleasure, though I sank into the moist, dark brown soil too deep for comfort, and found it very fatiguing to jump the quarter drains, small ditches at a distance of 200 feet apart, and, worse, to walk the very narrow plank over the 10-foot ditch which runs all around the field and is very deep.

The evening is beautiful; the sun, just sinking in a hazy, mellow light, is a fiery dark red, the air is fresh from the sea, only three miles to the east, the rice-field banks are gay with flowers, white and blue violets, blackberry blossoms, wistaria, and the lovely blue jessamine, which is as sweet as an orange blossom. Near the bridge two negro women are fishing, with great strings of fish beside them. The streams are full of Virginia perch, bream, and trout; you have only to drop your line in with a wriggling worm at the end, and keep silent, and you have fine sport. Then the men set their canes securely in the bank just before dark and leave then, and almost invariably find a fish ready for breakfast in the morn-

Near the bridge two negro women are fishing.

ing. There is a saying that one cannot starve in this country, and it is true.

As I drove down I saw little children with buckets and piggins picking blackberries; such big, sweet berries, covering acres of old fields which once were planted in corn. As I walked down the bank I found a "cooter" (terrapin) which had come out of the river to lay eggs. My excellent Chloe will make a delicious soup from it, or, still better, bake it in the shell. All winter we have quantities of English ducks in the rice-fields and partridges and snipe on the upland, and in the woods wild turkeys and deer, so that if there is a sportsman in the family, one can live royally with no expense.

Sheep live and thrive without any outlay. In 1890 I exchanged a very fine two-year-old grade Devon, for twenty sheep. Since then I have bought seven more. A gale, with sudden rise of water, destroyed twenty-two at one time in 1896, and I lost ten by dogs, but notwithstanding these losses, in the last seven years they have brought me in $200 by sale of mutton; my house is furnished with rugs and blankets, and I am dressed in serge made from their wool, and I have to-day at this place forty-six sheep and thirty-five splendid lambs. If I only could get the latter to a good market, it would pay handsomely, for their keep has cost nothing. I have a Page wire fence around my place.

In the same way cattle live and thrive with no grain, only straw during the winter, and the negroes do not give theirs even straw; they simply turn them into the woods, and in the spring look them up; find the cows with fine young calves and ready to be milked. They shut the calf up in a pen and turn the mother out, and she ranges the rich, grassy meadows during the day, but always returns to her calf at night. When she is milked, half of the milk is left for the calf. In this way the negroes raise a great many cattle, the head of every family owning a pair of oxen and one or two cows.

However, we cannot turn our cattle into the woods as we used to do, for unless we go to the expense of hiring a man to follow them, they will disappear, and no trace of them can be found. One negro will not testify in court against another, so that it is scarcely worth while to attempt to prosecute, for there is no chance of conviction. You hear that such a man has been seen driving off your animal; one or two people say they have seen him; you bring it into court, and witness after witness swears entire ignorance of the matter.

I, for instance, have 500 acres of pine land, and the family estate and my brothers' together make 3000 acres of the finest pasture land. Where my father had herds of splendid cattle I have to keep my cows in a very poor pasture of twenty acres, fenced in, and in consequence have only five or six cows and one pair of oxen on the same plantation where my father used to stable sixty pair of oxen during the winter. They worked the rice land in the spring and roamed the woods and grew fat in summer.

On the road this morning I met Wishy, who made many civil inquiries about my health. Five years ago one morning I was waked earlier than usual by a request from Wishy's mother, Annette, for something to stop bleeding. He had been badly cut by a negro, who struck him on the head with a lightwood bar. Wishy had laughed at his special flame, who had gone to church the Sunday before with a long white veil on her hat and he was enraged. I sent witchhazel and the simple remedies which I always keep for such calls. About eleven o'clock another request came, this time to lend my wagon and horses to carry Wishy to town fourteen miles away, as his head was still bleeding. I was shocked to hear that he was still losing blood and told them the drive might be fatal under the circumstances; I would go out and see what could be done.

Hastily getting together all the remedies I could think of, my niece and I drove to Annette's house, which was crowded with eager friends gazing at the unhappy Wishy, who sat in the middle of the room, leaning forward over a tub, a man on each side supporting him, while the blood literally spouted from his head, — not a steady flow but in jets. It was an awful sight. I had a bed made on the floor near the door and had him lifted to it, well propped up with pillows, so that he was in a sitting posture. At that time we had no doctor nearer than the town, except a man who had come from a neighboring state under a cloud of mystery. As soon as I heard of Wishy's condition I had sent for him, but

A request from Wishy's mother, Annette, for something to stop bleeding.

the boy returned, saying he was not able to read my note, so there was nothing but to do what I could or to let Wishy die.

I got Frank, who was very intelligent, to help me. I dipped absorbent cotton in brandy and then into powdered alum, and put it into the hole in the top of Wishy's head; it seemed a gulf! I put in more and more, having Frank hold his hands closely around the top of the head; but still the blood flowed. Then I sprinkled the powdered alum over all thickly until there was only one little round hole just in the middle; I made a little ball of cotton and alum and pressed it down into the hole with my finger and it was done. I gave him the milk I had carried, had the house cleared of people, and left, ordering that when the doctor came, I should be sent for.

A day passed, and when I sent milk, the message came back that the doctor had been there, looked at him, and gone away. I began to feel very unhappy over the heterogeneous contents of Wishy's head, but if I had not stopped the flow in some way, he would have been dead certainly — his pulse was just a flutter. I tried not to worry over it. The third day a runner came to say: "De docta' cum." With all speed I had Prue put in the buckboard and drove out. I had never seen the doctor and was surprised to find a fine-looking man in possession of the cabin. He called for a razor, said he could do nothing until he shaved Wishy's head. There was confusion among the numerous darkies who crowded round the house. At last it was agreed that Uncle Jack had the only razor in the street (as they call the negro quarters) that could cut. While a woman went for the razor, the doctor told Annette he must have hot water, and she proceeded to put a tomato can full of water on the fire; but he peremptorily ordered a large pot carefully washed, filled with water, and put on the fire. When the razor came, it was too dull to be of any use until the doctor had sharpened it, and then he shaved all of the woolly head.

I watched the man's proceedings with a growing feeling of shame. I had gone there to keep my eye on him, to prevent any roughness or carelessness to the patient, and he could not have been gentler or more interested and careful if he had been treating the Prince of Wales himself. It was a long business; with an endless stream of hot water from a fountain-syringe he removed from the hollow depths of Wishy's skull all the wonderful packing with which I had filled it, and I went away satisfied.

Day after day for three weeks he came and dressed the wound, until Wishy's head was restored to its normal state. Then he sent a bill for $20, which Wishy begged me to pay, and he would gradually return the money to me as he worked. Of course, I paid it, and, sad to say, not one dollar has ever been returned to me. Wishy married the next winter, and moved to a neighboring plantation. He has never even sent me a string of herring, though he has had a net for two years and caught great quantities which he sold readily at a cent apiece.

During the run of herring in the spring they crowd up the little streams in the most extraordinary way, just piling on top of each other in their haste to reach the very source of the stream, apparently. I suppose one little leader must wave its little tail and cry "excelsior" to the others. At a small bridge over a shallow creek near here a barrelful has been taken with a dip-net in an afternoon. But it takes a meditative, not to say an idle person, to watch for the special day and hour when the herring are seized by the impulse to ascend that particular stream.

I must stop now, not having said anything I meant to say, having been led away by the thought of my lost $20 and how very useful it would be to me now, and I will have to leave other things for another day.

PATIENCE PENNINGTON.

P.S. In future I will not write you a letter, but keep a diary and send you a few sheets from time to time. P. P.

PEACEVILLE, September 1.

Had a trying day at the plantation, making an effort to get hay properly stacked, and was detained late. I had told Jonadab to wash the buckboard and grease the wheels, which he had done very thoroughly, for I could hear the grease crackling, and Ruth was travelling very fast when the world seemed to come to an end.

I did not know what had happened, but flew to Ruth's head and quieted her, though she had dragged the buckboard some distance before I could stop her. I do not know what became of Dab at first, for I didn't see him until I had stopped Ruth, when he came up, stuttering fearfully, and said: —

"The wheel is lef' behind."

The front wheel had rolled off. I told him to go and bring it and put it on, though I did not see how he was to do it alone and I could not possibly help, as it was all I could do to hold Ruth. Jonadab, however, has a way of surprising me by unexpected capacity, just as a variety from my constant surprise over his awkwardness.

On this occasion he held the wheel in one hand while he lifted the axle with the other and got the wheel on. Then I sent him to look for the nut, but I felt it was a forlorn chance, for it was now quite dark. I was in despair; we were three and a half miles from Peaceville, and if I walked, I would have to leave all my impedimenta and only take my basket of keys and other small things, such as my diary.

Most of the planters go home at sunset, and I feared they had all passed, and I could not see my way to any solution. Just as I had come to the conclusion that even my resourceful mind could find no way out of the darkness, two buggies drove up and the gentlemen asked what they could do for me. I explained the situation, and one of them said: —

"If you will drive with me, Miss Pennington, Mr. B. will

take your things; the boy can ride the horse, and we will leave the buckboard here until to-morrow."

I accepted the hospitality of his buggy with many thanks. The transfer of freight was made. Dab took Ruth out, and I rolled the vehicle into the woods, as I could not bear that my buckboard should be left on the roadside, a spectacle of a breakdown. Just as it was all accomplished Dab stammered out: —

"I find de nut."

Great surprise, for this was fully 100 yards in front of the spot where the wheel had run off, but he said he felt it under his foot and picked it up and showed it in his hand. Mr. H. said: —

"That boy could never have put the nut on at all after greasing it!"

Dab was vociferous as to his having put it on and screwed it tight. I was beyond conjecture, and too thankful to question. Very rapidly the transfer was made back to my vehicle, Mr. H. remarking, "Your buckboard takes easily more than our two buggies."

I thanked them heartily for their chivalrous aid, and we all drove on home.

After the agitation had somewhat subsided, I asked Dab, who was sitting behind, if he had really put the tap on or not. He answered with great certainty: —

"Yes, ma'am; yes, ma'am; I did put it on; I know I did, en screwed it tight — "

I did not contradict him, but said, "Think about it; go back in your mind and remember just what you did, and where you put the nut when you took off this wheel, which you say you greased last."

After two miles in silence I heard convulsed sounds from the back, and finally out came "No-o-o, Miss Pashuns; no ma'am, I never put that nut on; I put it on the front o' the buckboard, an' when I put the wheel on, I went for a drink of water, and never did put the nut back — no, ma'am, I never put it back. I left it setting on the front of the buckboard."

When Dab finally gets started after stuttering and spluttering, he cannot bear to stop talking, but keeps repeating his statement over and over with delight at the glib way in which the words come out, and I have to say mildly, "That will do, Dab," and even then I hear him saying them over to himself.

After stopping his flow of speech I told him it was a great relief to know exactly what had happened, and I hoped it would be a lesson to him all his life and make him feel that he was a responsible being; that I trusted him with important work, and how if it had not been for God's great goodness, I might be lying on the road with a broken neck and he with both legs broken. I did all I could to make him feel what a mercy it had been, and he seemed deeply impressed. Ruth behaved beautifully during the whole thing, so I gave her a saucer of sugar when we got home.

Saturday, September 9.

Have been ill ever since the happy incident the other night, but this afternoon I felt impelled to go into the plantation. I had planned to send Chloe in with the money for Bonaparte to pay off, but at the last minute got up and went myself.

As soon as he saw me Green said: "Glad you cum, ma'am. Nana's got de colic turrible en we dunno w'at to do fur her."

I forgot that I was myself decrepit and flew to the house and got a bottle of colic cure and a box of axle-grease. I always keep aconite, but had none, but fortunately had not returned this bottle of horse medicine which I had borrowed when Ruth was sick. It said a teaspoonful every half hour, but I knew my time was short and Nana was desperately ill, so I gave a teaspoonful every ten minutes.

She would just throw her great body down with such force

that it seemed she must break every bone in it, roll over and over, and pick herself up and flop down again before you could attempt to head her off. While she was down, I made Green and Dab rub her heavily with axle-grease.

I myself put the medicine in her mouth, holding her head up and her lips tight together until she had swallowed it. She has such confidence in me that she did not resist at all, but kept quite still while I did it. I gave her six doses, and then it was dark, and I suddenly became aware that I was very tired and could do no more. I told Nana good-by, for I never expect to see her again.

My poor dear little Irish terrier, who is my shadow and constant companion, is very ill. For three days he has neither eaten nor drunk. His throat seems paralyzed, and he looks at me with such superhuman eyes that it makes me miserable, for I can do nothing for him.

I take a bowl of water to him and he buries his little nose in it, but cannot swallow or even snuff it up. I can get nothing down his throat, so that it is impossible to treat him.

Sunday.

Poor little Snap was so ill and made such a constant appeal to me for help which I could not give, that I felt it was cruel to let him suffer longer, so I sent to Miss Penelope for a bottle of chloroform. He followed me from room to room, a feeble skeleton, all eyes, and still I tried to give him milk, and when he turned his head from that, I gave him water into which he would feebly dip his little black-tipped mouth.

At last I took him in my arms and put him on a soft cushion in a tall banana box; then I cut several pieces of very savory roast beef and put them all around his little muzzle. He could not eat them, but he could smell them, and I could see by his eye that it was a comfort to him to have them there.

Then I filled a sponge with chloroform and put it into a cone which I had made of pasteboard and put it over his head and covered up the whole thing with a heavy rug. After two hours I sent Dab to look in, and he came back radiant to say that Snap was quite well.

I went to look, and the dear little doggie roused himself from a delightful nap to look at me. All expression of suffering and appeal was gone from his eyes. He looked supremely happy and comfortable, and after glancing up at me he tucked his head down on the roast beef and went to sleep again.

I wet the sponge and once more left him. When I took him out the next morning, I could not believe he was dead, so perfectly happy and natural did he look. Dab dug his grave in my little garden, and I laid him to rest, feeling the loneliest mortal on earth when I got through.

September.

When I went in to Cherokee yesterday, I was amazed to find Nana quite recovered. I had told Bonaparte if she showed any disposition to eat, to give her rough rice instead of either oats or corn, and it seems to have been a happy thought, for it agreed with her, and though weak still and much skinned and bruised by the way she threw herself about, she seemed quite well.

This is the eighty-eighth birthday of the sainted friend whom I visit every day. Every one in the little village sent her some little offering, so that her room was full of flowers and dainty trifles, and she enjoyed them so much. Though unable to eat anything and nearly blind, her interest in everything and everybody is vivid.

This afternoon, as Dab was putting the demijohn of milk in the box preparatory to leaving Cherokee, and I was standing in front of him screwing the top on the jar of cream to put in the same box, suddenly he dropped the demijohn and leaped in the air, uttering the most terrific Comanche yells I ever heard. I nearly dropped the jar of cream at the sound; he fled away still yelling.

My mind is fertile in horrors, and I said to myself, "The boy has gone mad!" I was terror struck.

When he finally stopped, some distance away, I called out, "What is the matter, Jonadab?" He just pointed to a spot near where I stood and began to yell again, "Snake run across my foot."

The relief was so great that I looked composedly on the big snake, but called in a tone of unwonted severity, "You must come and kill it." I knew the only thing to prevent Dab from going into a fit was to be severe in my tone, and peremptory.

Most reluctantly and slowly he returned. I cannot imagine why the snake elected to stay in the ivy to meet its fate; it was sluggish, evidently having swallowed something large, either a rat or another snake, for it was very stout. I made Dab find a long strong stick. It required continued urging and encouragement to get Dab to complete the job, but as soon as it was done and he felt himself victor over the thing which had terrified him so, he became puffed up with pride and courage.

September 30.

The storm is over, and all nature is smiling. Oh, the beauty of the sunshine falling on the dark green pines and the ecstasy of the song of the mocking-bird, who is perched on a tall pine just east of the piazza, splitting his little throat, trying to give vent to his joy and thanksgiving to the Great Father! If one could only bottle up a little of this sunshine and glory and ecstasy to bring out on some gray morning when one's blessings seem too far away to be remembered!

I am just writing a line while Dab is having his breakfast

and putting Ruth in the buckboard before we start for Cherokee to see the damage done by the winds and the deluge of rain which fell for twenty-four hours. The cotton had opened more fully Saturday than it yet had done, but a slight drizzle prevented its being picked. I fear the hay which was stacked will all have to be taken down.

8 P.M. — Spent the day at Cherokee fighting with incompetency and unwillingness.

The loose, irregular stacks of hay were, of course, wet to the heart, and I had them taken down entirely, much to Green's dismay. He thought it purely folly and fussiness, and I had to stand by and see it done, lending a helping hand now and then, to get it done at all.

He was loud in his abuse of Gibbey, his brother, for his incompetency and determination not to work, saying, "He's too strifflin' to lib," but that he himself was capable of everything; not only stacking hay, but everything else, he did in the most perfect way. I let him talk on, for his manner was respectful, and I was really interested and amused to see unveiled his opinion of himself.

It would be very comfortable to see one's self in that perfect light, instead of being always so fiercely conscious of one's own shortcomings. I almost envied Green his fool's paradise.

I went to a stack which he assured he had "'zamined, an' it was puffectly dry, 'cause, I put dat stack up myself." With ease I ran my hand in up to the elbow and brought out a handful of soaking wet hay. But that had no effect; he said that was some he had just thrown back, fearing to have it exposed, as it might rain, looking wisely at the clear sky.

One has to pray inwardly all the time to keep from a mighty outburst. He is better than any one else I could get just now.

Spent some time in the cotton-field seeing that the first pickings were spread on sheets in the sun so as to dry thoroughly. I had put some peanuts in my pockets for the little girls, Jean and Kitty, and I stayed talking to them a little while.

They have up to this time "minded child," that is, each has lived with a married sister and taken care of their babies. They do not look as though they had enjoyed life, nor have they learned anything, and I am anxious to brighten them up a little and teach them to take an interest and pride in their work. Thus far I cannot boast of my success, as to-day Jean picked six pounds and Kitty four!

October 1.

Another gorgeous autumn day, with just enough white clouds flying here and there to make shadows. The cowpeas were picked to-day, and they are bearing finely, and the people know how to pick them; it

Green thought it was folly and fussiness.

is not like the cotton. One woman who never can pick more than twenty pounds of cotton had seventy pounds of peas, and Eva had ninety pounds. I feel better satisfied with the day's work than usual.

I got the hay which had been dried put in the barn, which is much better than stacking it, when no one knows how, but I could only do that because the ground is too wet to run the mowing machine; thus I could use the team to haul in the hay. One of the renters came up and paid his money quite voluntarily, which is so unusual that it put me in good spirits for the day.

October 3.

To-day is too beautiful for words. As I went into the sun-swept piazza this morning I felt, like the mocking-birds, an ecstasy of gratitude for so much beauty. I did wish so I could take a day off and sit in the piazza and just bask in the beauty of everything and breathe the crisp freshness of the first fall weather and sew.

I am making a suit of white flannel woven from the wool of my own sheep. I have embroidered the revers and cuffs of the jacket and nearly finished it, and want it to wear these delightfully cool mornings, but I cannot stay to-day.

I must get through my home duties as quickly as possible and make my daily visit to the bedside of my saintly friend, who, having begun her life in wealth and having in middle age been reduced to poverty, has passed fourscore and eight years, a beautiful example of woman, wife, and mother, and is now slowly passing through the valley of the shadow. This is my greatest pleasure and privilege, and whatever other duty is hurried over, to this I give full time.

To hold daily converse with one who, after lying three months in hourly pain, is serene and calm, nay, joyous with gratitude for His many mercies (which some might need a microscope to discover), is a rare opportunity of seeing a true follower of the Blessed One, and I come away always feeling as though I had quenched my thirst at a living stream, refreshed and strengthened.

On the plantations, too, things look bright. The pea-vine hay is falling heavy and sweet behind the mowing machine, and what was cut yesterday has dried nicely and will be raked into windrows this afternoon. The crab-grass hay is also dry and ready to be stacked again. The cotton is opening well, and we can make a good picking to-morrow.

As I went into the pea-field, where the women were singing as they picked, I came upon a spider who was too large to stand upon a silver dollar. I was most reluctant to kill him, for he was doubtless the Hitachiyama of his race.

He scorned to run, or even move quickly away, so sure

was he that he was invincible and need fear no foe, and it did seem too unfair to crush out his little greatness, but the bite of such a spider would mean serious illness, if not death, and there were all the women, most of them with bare feet, to run the risk of being stung, so I dealt the fatal blow.

Some of the women picked ninety pounds, and Jean picked forty and Kitty thirty-six.

October 4.

Job knew what he was talking about when he said: "Man is born to trouble as the sparks fly upward." I went to Cherokee in quite an excitement this morning because the cotton-field was snowy yesterday and I expected to make a big picking, but last night, on a plantation three miles away, an old woman died and not a creature has come out to work.

Eva is the "Presidence of the Dessiety," her son tells me, to which Linette belonged, and so, of course, she could not be expected to work to-day, but the other women have no such eminence nor can they claim kin nor even friendship: meanwhile should the weather change and a rain come down, my precious cotton will be ruined.

October 5.

Another brilliant morning, but no hands in the cotton-field but Eva. She, having accomplished the duties falling on her as "presidence" of the burial society and pinked out yards and yards of frilling for the dressing of the coffin and shroud and sat up all last night, did not feel bound to remain to the funeral, as they had not been friends; indeed the departed Linette had been the cause of great domestic infelicity to Eva, so she came and picked her usual thirty-five pounds alone.

I sent Dab to pick for a short time, and he did very well, picking eleven pounds in about an hour. Then I went in and picked for about fifteen minutes myself.

I wanted to find out what the difficulty was. I picked

a pound and a half and found it very easy and interesting, even exciting work, and I am no wiser than I was before. If I was not afraid of the sun, I would have gone on all day, or rather until 2 o'clock, for it clouded up after that, and I came home in a pouring rain, which continues at bedtime.

October 6.

A beautiful bright Sunday after a night of heavy rain. The thought of the wasting cotton had to be sternly put aside. I had to visit the wonderful invalid before I could get rid of the nagging thought, "If only the cotton had been picked!" After that the glorious sunshine and singing birds had their full value, and the seventh-day peace reigned within as well as without.

I have a little class in the afternoon on my piazza for a Sunday lesson, eight little boys and one golden-haired, blue-eyed little girl. At first, I had some difficulty in getting them to come, for they always have such a good time playing that it seemed to them a great waste of the golden hours to come to Sunday-school.

Some of them said they were willing to come and sing hymns, but they did not want any lesson. However, I found one little fellow who wanted the lesson, so I told him to invite any one who wanted the lesson to come with him at 4.30 o'clock the next Sunday afternoon, but no one else.

Punctually at the hour three little boys and one little girl arrived, while the other boys in the village played up and down before my gate most ostentatiously, so that little heads could not help turning to see what was going on, and in the midst of one of the Commandments, I heard a squeaky little voice, "I wonder what those fellows are laughing at!" for they had got up a great burst out in the road, quite a stage laugh.

However, we got through comfortably and went into the

She picked her usual thirty-five pounds alone.

sitting-room to the piano, and I asked each one to choose a hymn which we sang. At the second hymn one of the boys from the road joined us, but I seemed unconscious of his presence, and when the singing was over, I invited the first four into the dining-room and handed them some little sponge cakes.

The next Sunday there was a full attendance and has been ever since. The lesson has to be carefully selected, as there are four denominations represented, so I take the Lambeth platform and teach the Creed, the Lord's Prayer, and the Ten Commandments. After that I tell them Bible stories, beginning with the thrilling narration of the Creation and the Garden of Eden.

When I first told how Eve was tempted to eat the fatal apple, and Adam too was tempted, and they were driven out from that beautiful spot to earn their living in the sweat of their brows, the interest was breathless, and one little fellow asked : —

"Miss Patience, what would have happened if they had never eaten the apple? Would they have stayed in the garden?"

"Yes," I said with confidence.

"And never had to wear any clothes?"

More faintly I answered "No, I suppose not."

"Well," he said, "the garden would have had to be made much bigger for all the children that were to come."

"Yes," I said, "I suppose the whole world would have been a garden," but I was glad to leave the subject and get on to firmer ground.

However, this Sunday when I asked them to tell me the story, they went on swimmingly until I asked who ate the apple first. Most chivalrously they all answered, "Adam."

"No," I said, "I am sorry to say it was Eve."

"Then," piped up the squeaky little voice, "then, Miss Patience, women are badder than men."

"Oh, no," I exclaimed, "but Eve was beguiled by the serpent, who told her the fruit would make her wise. The great Creator made man first, and meant him to be the protector and guide of the woman, and when she offered him the apple, he should have refused and said, 'Light of my eyes, we must not eat it. The Great Being who made us and gave us this beautiful home forbid us to eat of that fruit.' But Adam failed in his duty and ate the apple, and they were driven out."

My sturdy little brown-eyed thinker, who had been listening with profound attention, said: —

"Miss Patience, what would have happened if Eve had eat the apple and Adam hadn't?"

I was completely routed. "I cannot think what would have happened then."

There was a chorus of little voices: "Why, Eve would have been driven out, and he would have the garden for hisself."

I am quite sure when I was small we never asked such questions. Perhaps when it was read, as it used to be, in the Bible language, it did not take such hold on the mind as it does when narrated, but I am so eager to get their interest and attention that I tell them the stories instead of reading them, and with such success that nothing but force could keep them away.

Always have to light the lamp before we finish singing, but no one will give up his hymn, and as I read over each verse very slowly before we sing it, and they repeat it after me, it takes a good while. It is wonderful how quickly they learn the words.

One very small boy, who strayed in for the first time, when I told him he could choose a hymn asked for "Yankee Doodle," greatly to the amusement of those who had been coming two months. It is a pleasure to teach such bright children. At the end I always hand a few chocolates or some candy.

To-day the hands are "toting" the rice into the flats.

CASA BIANCA, October 8, 1903.

The harvest has come and with it real harvest weather — crisp, cool, clear; and the bowed heads of the golden grain glow in the sunshine. The hurricane which was reported as wandering around last week frightened me terribly, but after waiting Monday, Tuesday, and Wednesday for it to materialize, I had to cut on Thursday, for the rice

"You see a stack of rice approaching, and you perceive a pair of legs, or a skirt, as the case may be, peeping from beneath."

was full ripe, and though we have had some light showers, there has been no serious bad weather. To-day the hands are "toting" the rice into the flats.

You see a stack of rice approaching, and as it makes its way across the plank which bridges the big ditch, you perceive a pair of legs or a skirt, as the case may be, peeping from beneath. Men, women, and children all carry, what look like immense loads, on their heads, apparently without effort. This is the gayest week of the year. Thursday the field was cut down by the hands with small reap-hooks, the long golden heads being carefully laid on the tall stubble to dry until the next day, when it was tied into sheaves, which the negroes do very skilfully with a wisp of the rice itself. Saturday it was stacked in small cocks to dry through Sunday, and to-day it is being loaded into the flats, having had every advantage of weather.

If only no rain or wind comes until it is unloaded at Cherokee, fifteen miles up the river! I have sent for a tug to tow the two flats up on the flood-tide this evening — just now it is dead low water, and the flats are aground, which always scares me; for, if by any chance they get on a log or any inequality, they get badly strained and often leak and ruin the rice. Flats are one of the heavy expenses on a rice plantation — large, flat-bottomed boats from twenty to eighty feet long and from ten to twelve feet wide, propelled in the most primitive way by poles and steered by one huge oar at the stern. They can be loaded up very high if the rice is properly stowed.

I have sent to try and get some rice-birds for my dinner. These are the most delicious little morsels, so small one can easily eat six for breakfast, and a man makes nothing of a dozen for dinner. We used to get them in great abundance only a few years ago, but now the rice-bird industry has become so big a thing we find it very hard to get any at all. Formerly a planter hired bird minders, furnished powder and shot, and got several dozen birds from each one; but now the negro men go at night with blazing torches into the old rice-fields, which are densely grown up in water-grasses and reeds, the birds are blinded and dazed by the light, and as the fat little bodies sway about on the slender growth upon which they rest, they are easily caught, their necks wrung, and they are thrust into the sack which each man has tied in front of him. In this way a man sometimes gets a bushel by the time the reddening dawn brings him home, and he finds waiting for him on the shore buyers from the nearest town, who are ready to pay thirty cents a dozen for the birds, so that one or two nights of this sport give as much as a month's labor. Of course, it is hard to come out to cut rice the next day, so probably illness is pleaded as an excuse for his absence in the field.

This makes it more and more difficult to get the rice harvested; no one but one of African descent could spend his nights in the rice-field, where the air is heavy with the moist malaria, so it is his opportunity. The shooting of rice-birds has almost gone out, for the bird minders are so careless. They shoot into the rice and so destroy as much as the birds, almost; now blank cartridges are almost entirely used to scare the birds. Going round the field one day with Marcus, I said, with great relief: "I'm so glad not to see a single bird to-day." He laughed and said: "Miss, wait till de bird minders shoot." In a few seconds the bird minders became aware of my approach and up and fired very nearly at the same time. The birds rose in clouds so that the sun seemed darkened for a few seconds, and the noise of their wings was deafening. It seemed tantalizing not to be able to get any to eat. In spite of the tremendous report of the firing, it did no execution, for the old-fashioned muskets which are used have an enormous load of very coarse powder, but no shot.

Now, my flats are loaded, and I must start on my twelve-mile drive to the pine-land. As soon as I can have the flats unloaded I must send them back for the hands to harvest their rice. I do not pretend to overlook this. I try to put them on their mettle to do the best possible. Some respond, but the majority just poke along, doing as little as possible each day, so as to have longer time to strip the rice from the straw, and carry it home in bags, so that when it comes to mill, there is not enough to pay their rent. They know how I hate to take all they bring, I so like for them to have a nice little pile of their own to ship; it is very hard for me to believe what the foreman tells me, that they have been eating this rice for three weeks past.

October 16.

I have threshed the May rice, and it has turned out very well, considering the hard time it had for two months after it was planted. My wages field made twenty-five bushels to the acre and the hands nearly the same, only a little less, but it is good rice and weighs forty-six pounds to the bushel; and as I hear every one complaining of very light rice, I am thankful it is so good.

October 17.

I have had an offer of $1.05 for my rice in the rough, and I am going to take it, though I shall miss the cracked rice and the flour which we get when the rice is milled, and the rice will have to be bagged and sewed up, which is a great deal of work; but Mr. S. will pay for it at my mill, and that will relieve my anxiety about money.

October 18.

A hard day's work, but the sale has been most satisfactory, for as the standard weight per bushel for rice is forty-five, and my rice weighs forty-six or forty-seven, I have a good many more dollars than I had bushels, which is very cheering; and I have had grip and am greatly in need of cheering. Mr. S. weighed every sack and put down the weights and then added up the interminable lines of figures. I added them, too, but was thankful I did not have the responsibility, for they came out differently each time I went over them.

October 24.

The harvest of my June field (wages) began to-day. Though very weak and miserable from grip, I drove the twelve miles to Casa Bianca, and in a lovely white piqué suit went down on the bank. I timed myself to get there about 12 o'clock, and as I expected I met a procession of dusky young men and maidens coming out of the field. I greeted them with pleasant words and compliments on their nice appearance, as they all reserve their gayest, prettiest clothes for harvest, and I delight to see them in gay colors, and am careful to pay them the compliment of putting on something pretty myself, which they greatly appreciate. After "passing the time of day," as they call the ordinary polite greetings, I asked each : "How much have you cut?" "A quarter, Miss." "Well, turn right back and cut another quarter — why, surely, Tom, you are not content to leave the field with only a quarter cut! It is but a weakling who would do that!" And so on till I have turned them all back and so saved the day.

A field of twenty-six acres is hard to manage, and unless you can stir their pride and enthusiasm they may take a week over it. One tall, slender girl, a rich, dark brown, and graceful as a deer, whose name is Pallas, when I ask, "How much?" answers, "Three-quarters, Ma'am, an' I'm just goin' to get my break'us an' come back an' cut another quarter." That gives me something to praise, which is always such a pleasure. Then two more young girls have each cut a half acre, so I shame the men and urge them not to let themselves be outdone; and in a little while things are swimming. I break down some of the tops of the canes and make a seat on the bank, and as from time to time they come down to dip their tin buckets in the river to drink, I offer them a piece of candy and one or two biscuits, which I always carry in the very stout leather satchel in which I keep my time-books, etc.

Pallas.

Though the sun is fiery, I feel more cheerful than I have for a good while. The field of rice is fine, Marcus says, — "Miss, I put my flag on dat fiel'," — and insists it will make over forty bushels to the acre. I don't throw cold water on his enthusiasm, but I know it will not. However, the rice is tall, and the golden heads are long and thick. I count a few heads and find 200 grains on one or two, and am almost carried away with Marcus's hope, but will not allow myself to think how much it will make. One year this field put in the bank $1080, but I know it will not do that this year. There is no use to think of it.

I stayed on the bank until sunset to encourage the slow workers to finish their task. All the work in this section is based on what was the "task" in slavery times. That it was very moderate is proved by the fact that the smart, brisk workers can do two or three "tasks" in a day, but the lazy ones can never be persuaded to do more than one task, though they may finish it by 11 o'clock. I feel placid to-night, for half the field is cut down and will dry on the stubble all day to-morrow.

October 26.

Drove down to Casa Bianca as early as I could and found the hands cutting merrily. As soon as each one had cut a half acre they turned in and tied that cut on Saturday and stacked it in small cocks.

Again I am cheered and rested by the beauty around me. The sun is gorgeous, though the autumn haze is all over the wide expanse of level fields with every hue of green and gold. I get in the small patches of shade made by the tall canes and feast my eyes and thank the Great Artist who has made it all so beautiful.

The three flats are in position for loading to-morrow, the wind is still west, and so I hope the fair weather may last. My supply of candy and biscuits is much appreciated. I make my own lunch on the biscuits and a bottle of artesian water, which I always take with me. I would as soon think of eating snake's eggs as of drinking the river water, so full is it of animal life, I am sure. I know how it would look under a powerful microscope.

October 31.

Spent yesterday in the mill threshing out my rice, most trying to me of all the work, the dust is so terrible; but the mill worked well, and so did the hands — and better than all, the rice turned out well, thirty-five bushels to the acre, and good, heavy rice. So I felt rewarded for the dust and other trials. I was so determined to prevent stealing that I engaged the sheriff's constable to watch on the nights that the rice was stacked in the barnyard; and now that expense is over, and the pile is safe in the second story of the shipping barn. Next I have to thresh out the people's rice from Casa Bianca, which will be up in a day or two; then I will have a little time to have the upland crops seen after before the rice here, at Cherokee, which was planted very late, will be ready to cut.

Front porch — Casa Bianca.

CHEROKEE, November 4.

Yesterday I had my wages field of rice here cut. It is only eleven acres of very poor rice, which has cost a good deal of money, owing to the freshets. The only thing to be done now was to get it in with as little expense as possible, so I announced yesterday that it must be in the barnyard to-night. Bonaparte looked wise, smiled in a superior way, and said that was impossible — that perhaps by Tuesday it could be got in. I didn't dispute his wisdom or argue with him. I simply went into the field with the hands in the morning, yesterday, and stayed until it was all cut down. I told Bonaparte to put a watchman in the field, and left the choice to him. He said he would put Elihu; so I rested content until about 10 o'clock, when I began to get anxious about it. The best planter in my neighborhood had told me he had never known the stealing of rice so bad from the field. He attributed it to there being so little planted as high up the river on account of the freshet, so that rice is very scarce. This rice had not been good enough to warrant the expense of the constable, but I did not wish to lose the little that was there, so I determined to go over and see for myself. I called a negro boy of about sixteen years whom I had recently taken into my service, and asked him if he was afraid to row me over to the field. He hesitated and I went on: "I want to take some lightwood and a blanket over to Elihu, who is watching, for the night is very cold." At once he said he was not afraid at all, as the moon was bright. When I ran up to my room to get my wraps and my good Chloe found I was going, she said: "Miss Patience, le' me go wid you; I know well how fo' paddle boat, en yo ain't long git dat boy, en yu dun know ef 'e kin manige boat at night." Of course I was delighted to take Chloe; I sent Jake for lightwood, she took the blanket and I the matches. The getting in the boat was the darkest part, but once out on the river it was perfectly lovely — such a glorious night, the air so crisp and exhilarating. As we neared the field Chloe entreated me to be careful when I got out on the bank, for Elihu might take us for thieves and shoot; but I went very fearlessly, for I had a conviction that there was no Elihu there, and so it proved.

I told Jake to kindle a large fire in a sheltered corner of the bank, while Chloe and I walked all the way round the field. I can't describe the weird peace of the scene; and to make it more ghostlike Chloe insisted on speaking in a low whisper, as becoming the time and place, and reminding me that people from the next place might be hiding all around. No sign of any marauder, however, appeared, and I knew the fire on the bank would give the impression that I had installed my friend the constable, so I went back to the house entirely satisfied with the expedition. I charged Jake to say nothing on the subject to any one. Why will one try to exact the impossible? I lost my man, who has been with me fifteen years, this fall, and Jake is the substitute for the present.

To-day I stayed in the field again all day and succeeded in getting the rice tied and put in the flat by sunset. Then I said the flat must be taken up to the barn, but Bonaparte said that could not be done because there was "'gen tide." Of course all the men echoed that it was impossible, but I laughed at the idea, and climbing to the top of the rice, I sat there and told two of the young men to take the poles and push the flat out into the river — having privately asked old Ancrum who had stowed the flat if it was true that a flat could not go against the tide, and having heard from him that it was nonsense. The men pushed the flat out and poled it up the river with the greatest ease, and before dark it was safely staked under the flat house, so that my mind will be at rest about it to-morrow.

November 6.

Threshed out the rice to-day. It made only twenty bushels to the acre, and I hear rice has gone down very much. The hands now are whipping out the seed rice, which is a tedious business, but no planter in this county will use mill-threshed rice for seed. Mr. S., who bought my rice and who travels all over the South buying rice for a mill in North

Carolina, told me that everywhere else mill-threshed rice was used, simply putting a little more to the acre. Here it is thought the mill breaks the rice too much, so the seed rice is prepared by each hand taking a single sheaf at a time and whipping it over a log, or a smooth board set up, until all the rice comes off. Then the sheaves are laid on a clay floor and beaten with flails, until nearly every grain has left the straw. After all this trouble of course it brings a good price — $1.75, $1.50 per bushel, $1.25 being the very cheapest to be had.

November 7.

The time for paying the taxes will soon be passed, and all the negroes on the place have asked me to pay their taxes in addition to my own, so that I must sell some rice. Took samples to our county town; I was told they were very good rice, but no one wished to buy. I was offered, however, $82\frac{1}{2}$ cents a bushel for one and 85 cents for the other! I sold the smaller lot for $82\frac{1}{2}$ and determined to hold the larger part, for I feel confident rice must go up by February, and I do so want to get $1 a bushel for it, for then I will pay out, but otherwise not, after all my work.

November 12.

Peaceville has been wrought up to a state of wild excitement. On Sunday afternoon, when I was expecting my little class, only Kitty and the Philosopher and Squeaky came, and before I could ask where the others were they burst out: —

"All the others have gone to hear the lion roar, and to see if they could get a peep at him."

"A lion? Here?" My tone was suitable to the subject.

"Yes, ma'am; they put up three big tents while we were in church this morning, right in front of the post-office."

I praised them for coming under such heavy temptations, but they exclaimed in chorus: "We didn't want to come — mamma made us; we wanted to hear the lion roar, too."

At which I was more pleased than ever, and was as rapid as possible with the lessons and told no story, though I thought Daniel in the lions' den might suit the occasion; but I soon saw that they could listen to nothing under such phenomenal circumstances. A very feeble Punch and Judy is the greatest show seen here before.

We sang the hymns, I gave each one an apple, and said I would walk down with them to the tents. A most delightful progress we made, every one having turned out to see the unwonted sight.

Before we got to my gate the King of the Forest began to roar tremendously and kept it up, to the awe and delight of the humans and the dismay of the animals. Cows refused to come up to be milked, but fled to the swamp, and horses cowered in their stalls.

Every one, even the most sedate, had turned out to look at the tents. I went with the children until I saw their parents and then returned to my piazza.

Tuesday.

Yesterday was the grand day. There were two exhibitions, one at 1 o'clock and at 8 P.M. The two stores were shut for the day, and business suspended while the village gave itself up to dissipation.

I had to go to the plantation, having an appointment with a carpenter for an important bit of work. It was difficult to get Ruth past the tents. I took the plan of stopping to talk to every one I met as I approached the green in front of the post-office, which was so changed since Saturday, when she saw it last.

Most fortunately the lion did not roar at that time, and we got by without accident. Though I have seen a great many fine wild beasts, the excitement in the air gained me, and I was anxious for Chloe to choose the morning performance as I had to be away then; but Chloe, when I told her she could go morning or evening, whichever she preferred, said she would go at night, as she heard that would be the grandest. So I could not go, for she would never have consented to leave the house and yard unguarded.

I did not see the show, but I certainly have enjoyed the accounts of it and have come to the conclusion that the Shelby show might be called a high-class moral entertainment. The most particular and sedate, not to say prudish, were not shocked, and the acrobatic feats amazed every one.

Peaceville was a great surprise to them also; they asked for a hotel or boarding-house; there was none. They wanted to board somewhere, but no one took boarders. The acrobatic star, who, as Chloe described her, hung from the top of the tent, dressed in "pink titers," by one foot, holding up her fifteen-year-old daughter, also beautiful in pink tights, by the foot, said she did not wish to stay in a tent; she never did; she wanted to be in a house, and finally some ladies who lived near the place where the tents were pitched said they had an empty house in their yard which they would fix for her, and it being Sunday afternoon and no servants were to be found, the ladies themselves put beds in the house and made it comfortable for the acrobat ladies, and when these offered to pay, were quite shocked and surprised and said there was no charge; they were glad to have been able to make them comfortable.

Chloe and Dab have both given me thrilling accounts of the lady dressed in pure silver, a very stout lady who took the head of a snake, bigger round than Dab's body, and stroked it and laid it on her breast: "Her color was quite change while she did it, en the snake lick out 'e tongue en you could see the lady trimble an' it was byutiful."

Altogether for many days joy will reign in the memory of these delights. It was conducted with great dignity, and there was no confusion or trouble, which seems wonderful, for there were great crowds of darkies coming from miles around and only about thirty white people all together. Yet they had the seats arranged on different sides, so that all were satisfied. The lion was given part of a kid before the spectators, and then he stopped roaring.

November 18.

Green has returned to work; that is, he milked this morning and hauled one load of manure to the field. His cousin, Wishy, got his kinfolk to buy off the negro who was prosecuting him for killing his cow, and the case was dropped.

Long ago, when I kept Wishy from bleeding to death by patching up his head, I fear I did not benefit the world.

I find Elihu has gone! Moved bag and baggage to my neighbor's, where he will have unlimited credit. He owes me $10, which he promised faithfully to pay, and Jean and Kitty have walked off in my boots beyond the reach of my small efforts to improve them.

I feel quite sad about it — my heart has always been tender to Elihu; I have had to help him so often. The last time he went off to make "big money," as they call it, on some timber work he came back very ill, and for a month I took him nourishment and medicine daily, in spite of which his wife and children lived in my potato patch. He was very weak, and one day he broke out: "Miss, if I ever lef' you 'gen and gone off for work any ways else, you sen' for the sheriff en tie me. You ben good to me en ten' me, en den de debil mek me lef' yu fer mek' big money! en now look a' me! Yu ten' me en yu feed me des de same."

He is an uncommonly rich shade of black, so that his own mother always referred to him as "dat black nigger." Under constant and proper supervision he can be very useful, but he cannot make himself work every day. He must have a compelling hand and head behind him.

He has ten living children and a smart active young woman

for his second wife. When we were planting largely of rice, he made a fine living, as he rented sixteen acres — he did the ploughing and his family the rest of the work. He had a splendid yoke of oxen, which he bought from us, and cows and another fine steer he had raised.

The changes in the conditions in the last few years I do not understand, but since McKinley's death steadily the negroes have declined in their responsibility and willingness to work until now their energies are spent in seeing how little they can do and still appear to work so as to secure a day's pay.

Elihu used to be a splendid ploughman, but this spring I had him to plough ten acres for me, breaking it up flush. The earth was barely scratched, I found afterward, though I paid him by the day instead of by the acre, fearing he would be tempted to hurry over it if I paid by the acre.

Forage was very scarce, and as long as he ploughed for me I told him to give his oxen all they could eat from the hay under the barn which was blown down. The two-story barn was packed full of hay, some of my best alfalfa, when the storm struck it. Of course it took some labor to get the hay out, and poor Elihu, after the mighty effort of ploughing one-half acre a day, could not make himself get out more than just enough to keep the oxen alive.

I had urged him from the beginning of the winter to make his children gather daily a certain quantity of the gray moss with which the oaks are laden and which cattle eat greedily; that would have kept his cows and oxen in good condition, but he never did it.

I had two large sacks gathered every day for my cattle; his went hungry. One by one the cows and young calves died, not being accustomed to range like the woods cattle.

Some time after he finished ploughing for me he drove his son up to see a doctor fifteen miles from here in a very bitter spell of weather — drove the creatures up without feed, and after consulting the doctor turned right back. One ox dropped and died two miles from home, the other managed to get back, but lay down about 100 yards from my front gate, under the trees laden with food which would have saved its life, if given in time. I used to take the lantern and go out at night to carry food to it, knowing that if Elihu saw me feeding it he would cease giving the little care which he expended on it.

It struggled on a week and then died. One month before he had been offered $60 for the yoke.

At last he had not an animal left. Then he came to me and said he would like to take service with my neighbor by the month as ploughman, as he would no longer give him work unless he hired to him by the month. I was very sorry, for I let him work there all the time when I had no work for him. He is a splendid boatman, and I always called on him to row me across the river and did not mind wind or waves with Elihu at the oars.

However, I told him he could do so if he paid $1 a month for his house — now he has gone, owing me for eight months rent besides his tax. Several years ago he was double taxed, having neglected to pay at the right time, and since then I have always paid his tax when I paid my own.

He owns some land with timber. When I went to pay the tax, I saw two buildings and twenty-five acres and the tax was $4. I saw Elihu, I showed him the paper, and asked: —

"Have you any buildings on the land?"

"No, miss, I ent build no house, I ruther stay here, en if I sick you ten' me."

"But, Elihu, the tax paper calls for two houses."

"Well, miss, ent you know, look like I ought to had house by now!"

"But if you have none, you should not pay tax on one.

Now when February comes, which is the month to make returns, I will make your return without the house."

"Well, miss, if you tink so, but I hate fer tek off de house."

I deprived him of his air castle, but the tax was reduced to $2.70, I believe — I must look over the tax receipts to see.

I always pay Bonaparte's and some others, I am so afraid of their putting off until they are double taxed. I do not see how I am to pay my own taxes this year; they are nearly $200,

Elihu was a splendid boatman.

and there is nothing coming in. I have many, many valuable things which I would like to sell, but I have no gift that way.

After many struggles I made up my mind to accept an offer for my castle in the air, a mountain top in the Sapphire region of North Carolina, but the purchaser withdrew; it is so with everything — no one wants to buy anything   If our valiant, voracious, and vivacious King Stork would only desist from his activities while a few small creatures were left it would be a mercy; but I fear when he gets through, there will be none but sharks, devil-fish, and swordfish left.

November 20, Saturday.

When Green came this morning, I told him I wanted Bonaparte to sow the oats on the land he has been ploughing this week, and he must harrow it in to-day, as the season is already late. He seemed shocked and said the land was quite too rough for him to get through harrowing the acre and a half to-day.

I in turn was shocked and told him that was absurd and that it must be done; that I was distressed to hear he had ploughed it so badly as what he said would indicate; that I would have Dab take Romola and run the cultivator while he ran the harrow, so as to have the oats thoroughly covered. I told Dab to get the horse at once and take the cultivator to the field.

I did a thousand things before following him. I found him in the slough of despond and I had to fix the harness, etc., for him, and then we proceeded to the field. I found Dab had not the faintest idea of how to guide the horse and manage the cultivator, so I told him until he got accustomed to it I would lead Romola, so that he could devote all his attention to the cultivator.

The ground was rough to distraction, and with every polite intention Romola could not help every now and then walking up my skirt, short as it was, and I was nearly dragged down upon the ground, but I could not bear to give up, though I was utterly exhausted, for the cultivator was doing good work.

We had just got through half an acre and I was wondering how I could retreat with my laurels, when Patty came at a full run to say the "lady had come." Never was an arrival more welcome. I told Dab he must take Romola back to the stable and make himself presentable and bring in dinner as soon as possible.

Made my way to the house as quickly as I could, but I was so tired that my feet were like lead. S— was very much surprised to find what I had been doing and proceeded to argue with me, but I only made fun of her arguments, and we had a very gay dinner.

My little brown maid Patty is a new acquisition and a great comfort, for she is very bright and intelligent and not too dignified to run, which is a great blessing.

CHEROKEE, Sunday, November 22.

Drove S— to church in our little pine-land village; she seemed to enjoy the very simple service. Then I took her over to my summer-house which is just across the road from the church. She was amused at the roughness and plainness of the pine-land house as compared to the winter quarters. Drove her then in to Hasty Point, which is named from Marion's hasty escape in a small boat from the British officers during the Revolution, and is a very beautiful point, overlooking the bold Thoroughfare and Peedee River; then home to a dinner of English ducks. I am very stiff from my agricultural efforts.

My little brown maid Patty is a new acquisition and a great comfort, for she is very bright.

November 24.

Yesterday just as I was getting into the buckboard to drive S— down to Gregory to take the train Jim arrived. He has come to begin the colts' education and can only stay a month, as his employer in Gregory gave him a month's holiday. I am so glad to have him — told him to get all the harness together and mend things up and see if he could contrive a harness fit to put on Marietta to break her in the road cart.

S— was so anxious to see Casa Bianca that I thought we could drive in there on our way to Gregory, eat our lunch there, and still get down in time for the train, but we failed to do it. She was so delighted with the place and wanted to see everything in the rambling old house, even the garret with its ghostly old oil portrait of a whole family in a row and a broken bust of another member, that we delayed too long. Besides, the train left at 4:10 instead of 4:45, as it has been doing for some years. I had to leave S— to spend the night at the hotel, which I hated to do, but she said she must get

The roughness and plainness of the pine-land house.

off on the 6 A.M. train, and I was equally obliged to come home, so we parted with mutual regret.

It was late for my long, lonely drive. By the time I got

to the ferry it was dark, and I wondered how I was to manage. I asked the two old men to lend me their lantern, but they said they could not spare it. However, about half a mile farther on I stopped at a cottage and asked for the loan of a lantern, and the owner, a darky, brought out a bright, well-trimmed lantern and with true courtesy assured me he was happy to lend it, and I made the drive without accident, truly thankful to get into my dear home, with its bright fire of live-oak logs, at 8 : 30 out of the cold and darkness.

December 8.

To-day Richard Dinny came to say he would undertake to mend the break in the rice-field bank. As it is about two miles round there in a boat, I had him paddle me through the canal to Long field trunk, and I walked from there on the banks. I hurried along because the time was short before hour for luncheon. I had had the bank hoed just in the middle, so that a sportsman could go through unseen by the ducks in the field. Sometimes it was hard for me to get through with my skirt, but the man found it hard to keep up with me. The break looked very alarming, the water rushing over, and every tide that goes over will double the work.

Coming back, my hair caught in a brier and I found it impossible to disentangle it. I had taken off my big hat early in the engagement and left it on the bank near the boat. After trying desperately to get free from the brier I asked Richard, who was just behind, if he had a knife. He said yes.

"Then cut this bramble," I said, holding well up above my head the brier, which was completely wrapped in my hair.

He got out his knife and took a long time about it, sawing and sawing, but finally I was released. As soon as I got home I rushed upstairs to fix my hair for luncheon, for it is curly and was every which way over my head. As I took it down a lock as thick as my finger came off in my hand. Richard had taken so long because he was sawing off my hair instead of the bramble.

December 9.

Yesterday's work at the break was too much for Richard. This morning he sent word he was called off by important business, so could not come.

December 11.

We are having the most delightful springlike weather. It is a joy to wake up morning after morning and find the same balmy, mild air. The effort to keep the house warm in the cold weather got on my nerves very much, and now I am relaxing and expanding to my own natural condition, which is rather optimistic — one of peace and good-will to the world in general, with a firm faith that things must come right in the end, however difficult and crisscross they may seem.

Went to Casa Bianca to-day. The place is too lovely for words. How any one who has the money and wants a winter home can hesitate to give $10,000 for it I do not see. When it is sold, it will break my heart, but either this place or that must go. This place (Cherokee) has nearly 900 acres, and the house is in perfect order. Besides, it has an ever-flowing artesian well 460 feet deep which throws water above the roof when a smaller pipe is put on, — a reducer, the man who bored the well called it. There is a grove of live oak of about 50 acres.

I often wonder that it should have fallen to my lot to have two such beautiful homes. Altogether if I only had a small certain income, I would not envy the King on his throne.

December 12.

All the sashes up this lovely April morning. I have a man called Jimmie trimming up a little. The vista my dear mother had cut out years ago had grown up, and it is a great pleasure to have it open once more. From the front piazza it opens a view down the river, a beautiful bend, the shining, glimmering water framed by the dark oak branches.

Finally I have put Joe, Ruben, and George to work on the break. After lunch went over in the boat to see their work; they had a fine supply of mud cut, some on the bank and some in the flat. Sent Bonaparte to take over some long plank for them to use inside of puncheons to hold the soft mud.

December 13.

Joe, George, and Ruben working on break. They had to be there at daybreak to catch the low tide. This afternoon I went over in boat to look at the work, and to my delight it is really done, and I believe will last, only every day at low water they must put on a little fresh mud to raise it as it settles.

Oh, this heavenly Indian summer! It is too delightful for words!

Bonaparte had Frankie and Green helping him to clean the chimneys. It is a troublesome business.

Bonaparte goes up on a ladder to the top of the house. It always frightens me to see him, for he is an old man, but he minds it less than the younger ones. He ties a stout cedar bough to a long rope about midway in the rope, then drops it down the chimney the three stories to the first floor; there Frankie catches the rope and between them they pull it backward and forward until the chimney is clean and the hearth is filled with soot.

Once I tried getting a chimney-sweep, but he wept and pleaded so not to go up the chimneys again, saying he would suffocate, they were so long, that I returned to the old and primitive way and will never try the sweep again. After this one sweeping we keep the chimneys clean by burning them, when there is a pouring rain, about once a month.

I have always broken my colts myself; no one but myself

either rode or drove Ruth until she was thoroughly broken. Of course Jim's stable discipline was of the utmost importance, and he always went along, but he never touched the reins. I did the driving.

This year, however, I had not the spirit to cope with them and have determined to leave it entirely to him. He is now patching up a harness so as to drive Marietta in the road cart.

NOTE

It may be wise to explain a peculiarity of our low-country rice region. From the last week in May until the first week in November it was considered deadly for an Anglo-Saxon to breathe the night air on a rice plantation; the fatal high bilious fever of the past was regarded as a certain consequence, while the African and his descendants were immune. Hence every rice planter had a summer home either in the mountains, or on the seashore, or in the belt of pine woods a few miles from the river, where perfect health was found. In 1845 my father built a large, airy house surrounded with wide piazzas on Pawley's Island, and there he spent the summer, with occasional trips north and abroad, until the war made it unsafe to occupy the island. Then he built a log house in the pineland village of Peaceville: this large house with double shingled roof was built by his plantation carpenters with wooden pins, owing to the blockade there being no nails to be had. After the war my brother owned this, and my mother in spite of great difficulties returned to the beach as a summer home. As the crow flies this island was about three miles east of Cherokee, but for us mortals to reach it, many miles by land and water had to be traversed — all of our belongings, servants, horses, cows, furniture, were loaded on to lighters and propelled seven miles through broad rivers and winding creeks to Waverly Mills where they were disembarked and travelled four miles by land, but when we reached this paradise on the Atlantic Ocean we felt repaid for all the effort. It was here we spent our summers when I began my rice-planting venture. As my mother reached the limit which David places for the span of life, she shrank from the long move and bought a house in Peaceville just opposite the church and here the last beautiful summers of her life were passed in peaceful serenity.

CHAPTER II

January 1.

ON the rice plantation the first of January is the time for the yearly powwow, which the negroes regard as a necessary function. It is always a trial to me, for I never know what may turn up, and the talk requires great tact and patience on my part, not more, I suppose, however, than any other New Year's reception. One is so apt to forget that the "patte de velours" which every one uses in polite society is even more of a help in dealing with the most ignorant, and makes life easier to all parties.

Saturday, January 2.

I went down to Casa Bianca for the important talk. I found two more families had been seized by the town fever. Every year more hands leave the plantations and flock to the town, and every year more funerals wend their slow way from the town to the country; for though they all want to live in town, none is so poor but his ashes must be taken "home"; that is, to the old plantation where his parents and grandparents lived and died and lie waiting the final summons. I met such a procession to-day, an ox-cart bearing the long wooden box, containing the coffin, and sitting on top of it the chief women mourners, veiled in crêpe; behind, one or two buggies, each containing more people than it was intended to carry; then behind that a long, straggling line of friends on foot, all wearing either black or white, for their taste forbids the wearing of any color at a funeral. The expense of a railroad journey does not deter them from bringing their dead "home." The whole family unite and "trow

The yearly pow-wow at Casa Bianca.

een" to make up the sum necessary to bring the wanderer home, and even the most careless and indifferent of the former owners respect the feeling and consent to have those who have been working elsewhere for years, and who perhaps left them in the lurch on some trying occasion, laid to rest in the vine-covered graveyard on the old plantation.

Two years ago, a man and his wife, of whom I thought a great deal, who had been married and who had lived always at Casa Bianca, left me to go to town. They had prospered and bought the usual progression — oxen, cows, a horse, and finally a house and lot in the county town, Gregory. This house they rented out for several years, and then the desire came to go and occupy and enjoy the house and give up the laborious rice planting. It seemed very natural, and though I was very sorry to part with them I could not say a word against the plan. Dan and Di were both splendid specimens of physical health and far above the average in intelligence, capacity, and fidelity. They went well provided, according to their standards. With his horse and wagon Dan supported his family in comfort, hauling wood, etc., while Di opened a little shop in one of her front rooms, which was well patronized, as their house was on the outskirts and far from the shopping street.

One afternoon, some months after their move, Di said to Dan: "I'm dat sleepy I haf tu lay down, but call me sho' befo' de sun set." She left Dan smoking his pipe on the little porch, where, about an hour later, the youngest child came to him for something, and he said, "Go ax yo' Ma, 'e toll me tu wake um fo' de sun go down." The baby went and returned reporting, "Ma 'oudn't answer me." Dan went in to find her dead. He brought her home to the plantation, and in a few months his son brought him also, to rest under the moss-laden live oaks.

This is only one instance out of many; those accustomed

to regular outdoor work cannot stand the confinement and relaxation of town life.

But back to the powwow at Casa Bianca. The two families who are moving to town carry off four young girls who are splendid workers, and very necessary to the cultivation of my "wages fields." Two of the men announce they are tired of renting and want to go "on contraak." This I do not quite understand, as they always sign a paper promising to do all that is required on the place, which I have considered a contract; and I am a good deal amused over their efforts to explain, when at last Marcus, the foreman, says to them: "De lady aint onde'stan', kase he neber wuk contrak, but I will make she sensible," which he proceeded to do with great delicacy. I found it simply was to work entirely for wages and not rent,

"Four young girls who are splendid workers."

and I was expected to give each one a half acre of rice land to plant, in addition to their house and large garden free of rent, in return for which they were to sign "contraak." It is impossible to show by the writing the funny emphasis which they put on the last syllable of this word.

The two hands were poor renters, so that the present arrangement is perfectly satisfactory to me, only the portion of land rented grows smaller year by year, and where is it to end? I cannot plant more land on wages than I do, for it costs $15 per acre, besides the keeping of the banks and trunks on the whole 200 acres. Last year there were ten acres less than the year before, and this year there will be twenty-five acres less than in 1903. Besides this, the plantation to the north of Casa Bianca, whose lands adjoin, has been practically abandoned, so that the water rushes down through its broken river bank on my fields, and I have to go to a heavy outlay to keep it out.

Marcus asked me to go round the bank with him, and after thinking it well over I have concluded to throw out three of my fields and make up a straight bank from the upland down to the Black River, a distance of half a mile, high, wide, and strong enough to act as a river bank, and resist the rushing water which comes with immense force in the Black River, for it is the deepest stream in this section, in many places 60 or 70 feet deep. It will cost a lot, and I do not know where the money is to come from; but if I do not make the stand against the water, I shall not be able to plant anything, and this is the place from which I derive my income.

"He that regardeth the clouds shall neither sow nor reap." This text is my great stand-by when things look stormy and I am discouraged. I suppose the rushing river may be considered as in some sort a relation, or at any rate a remote descendant, of the clouds, and I will not regard it, but give Marcus an order to go to work on the bank as soon as possible.

The week after this visit I was sent for by the foreman at Casa Bianca. When I went down, I found every one in a state of unrest and ferment. Nat, one of the renters, had told the others that he had made a special arrangement with me by which he was to do only what he wished to do. Now, one would suppose that no sane person would believe such a statement as this, but I had been seen talking to Nat apart, and they were all prepared to throw up their agreement and go — "contraak" hands and all. It was some time before I found out what the matter was, for even Marcus was entirely upset and talked mysterious nonsense before he finally gave me the key to the situation. I then assembled all the men and told Nat to recount what he had said to me on that occasion and what I had said to him. He pretended to have forgotten. So I related: "You told me your mother wanted to move to town and take your three sisters with her, so that your working force would be diminished, and you would not be able to rent as much this year as you had done, and you would want only eight instead of twelve acres. I told you I was sorry your mother was going, for though she herself no longer worked the girls were good hands. Then I asked you if you remembered when your mother first came to me. You were a very little boy; she was in great distress, having been turned away from the place where she was living, with her large family of little children. All her things had been put out in the road because she had been fighting, and she entreated me to give her a house to stay in. I told her I heard that she was a 'mighty warrior' and stirred up strife wherever she went. But she promised not to 'war' any more, so I gave her a house and she kept her promise, prospered, brought up her large family respectably, and now owned much 'proppity,' cows, oxen, and pigs, and everything she wanted; and the children had all grown up healthy and happy, and I only hoped they might retain their health of soul and body in town."

They all listened attentively while Nat punctuated my narration with "Yes, ma'am," at every comma. Then I said: "Did I say anything more to you, Nat?" "No, ma'am; dat's all." Then indignation broke out on Nat from the assembled hands. "En yo' tole all dem lie fo' mek we fool! I mos' bin gone way," and much more, all talking at once. Nat only looked foolish and said: "I jes' bin a fun." I gave him a serious talk, and the hands scattered in high good humor; but if I had not gone down that day, in all probability the whole party would have packed up their household goods in their ox-carts and left, "contraak" hands and all! Marcus said, with his usual dignity: "Myself, ma'am, bin most turn stupid" — as though no words could express more fully the seriousness of the situation.

March 12.

Since then things have gone very comfortably and quietly at Casa Bianca. The field I am to plant in April has been well ploughed, the ditches cleaned, and finally the division bank made up splendidly. Across the canal and down to the river it is one foot above "full moon tide." I have twenty-six fine lambs, born in January. At Cherokee, also, I have had some good work done. My "wages field" there was ploughed early in February, so that the frost has had a chance to mellow it. I have ten acres of fine oats growing and ten acres prepared for corn; pigs, cows, and everything doing well, except the lambs. Nine were born in January, but some "varmint," Bonaparte reports, has killed seven. I know the "varmint" is a dog, somebody's treasure, so that it cannot be convicted, and every other animal is suspected, — fox, wildcat; and many strange tracks are talked of.

She promised not to war any more.

F

At Cherokee I had to put down a new trunk, which is quite a business. It requires knowledge of a certain kind, but is very simple, like most things, to those who know. To me it seems a terrible undertaking, for if it is badly done, the trunk may blow out when the field is planted, and ruin the crop. Knowing so little as I do, I thought it best to leave it to Bonaparte, so I did not go over to the place, which is about a mile away through winding creeks.

The tide suited the morning, January 12, and the weather was mild. I waited with great anxiety for the return of the hands in the evening. I rushed down to the barnyard when I heard the boat, and asked if the trunk was well down. Bonaparte smiled in his superior way.

"Well, no, ma'am; de fac' is we neber did git de ole trunk out."

"What," I said, "you have left it half done?"

"Oh, no, ma'am. We bruk up de ole trunk an' tuk out all but de bottom."

"Then the water is rushing through to-night, making the gulf wider and wider?"

"Yes, ma'am."

"Myself, ma'am, bin most stupid."

I was speechless. There was no use saying anything, but I decided to go over the next day and use my common sense, if I had no knowledge. Bonaparte told me he could not get the hands, any of them, to go down in the water, and no trunk can be buried with dry feet.

The next morning, January 13, I went, carrying lunch and a bottle of home-made wine, with a stick in it for those who were to get wet. It was a beautiful bright day, with the thermometer at 50 at 9 o'clock, for which I was very thankful. The tide was not low enough for anything to be done until then. I had had three flatloads of mud cut and put on the bank, and everything was at hand. The getting up of the bottom planks was at last accomplished, and then the new trunk floated in to place on the last of the ebb, so that it settled itself into its new bed on the low water, and then the filling up was a perfect race, so much mud to be put in before the tide began to rise, besides the inclination of the bank to cave in. I kept urging the two men down in the gulf to pack the fresh mud well as it was thrown in. The stringpiece and ground logs which Bonaparte had provided were, according to my ideas, entirely inadequate, and I sent four hands to an island near by to cut larger, heavier pieces. Altogether, the day was one of the most exciting and interesting I ever spent, though I stood six hours on the top of a pile of mud on a small piece of plank, where I had to balance myself with care to look into the gulf and not topple over. It was black dark when we left the trunk, but the mud was well packed, with every appearance of solidity and stability, and the next day I had two more flatloads of mud put on, and, though a freshet has come and gone since, "she" has not stirred, and the field drains beautifully.

The company which planted the places next to Cherokee has broken up. One of the principal investors told me that he had had his money in it for seven years, and never got a cent of interest, and he was thankful to get out of it. They have taken all my best hands, one by one, but they have not succeeded — did not make money for all that. And this year the price of rice has gone down, so that what has been made brings only half of what was hoped for.

I believe these lands would make a great deal if we understood the cutting and curing of hay, for the grass grows most luxuriantly if the land is ploughed and left, but the curing of hay is unknown to the rice-field darky.

Our uplands are very fertile and adapted to any crop, it seems. Last year a few tried cotton and did very well without any commercial fertilizer. The only trouble is the nice cultivation cotton requires. I had a little planted. We did very well and sold at 15½, but I am afraid to plant more than an acre or two, as I cannot get it kept clean, which is essential to cotton.

The freshet, which we always look for after the melting of the snows in the mountains, has not yet come. We had one a month ago, but now it has subsided, and the rush of preparation for planting should go on; but I find it impossible to enthuse my renters. A lethargy seems to have fallen upon them, and if I only had the money, I would plant all the available land myself. But that is a very big if, and I must just have patience and try to rouse their energy. Above me there are only a very few acres planted, as the freshet is more disastrous the higher up the river you go. About two miles above me is a historic plantation, where Marion made a very narrow escape from his British pursuers by jumping into a canoe and pushing up a small creek, while the British, after some delay in getting a boat, rowed, as they thought, after him, but followed the bold, wide stream of the Thoroughfare, which took them rapidly away from him.

This is the home of a very remarkable woman, who has, by her own exertions, educated her sisters and brothers and paid off the mortgage on the plantation. The family was wealthy and accustomed to the liberal use of money, but when the end of the war came, they found themselves with nothing but the land, not a cent to plant or to buy food. This young girl received a present of a small sum of money from a relative in England, which she invested in supplies that every one was in need of, opened a small store, and as fast as she sold out reinvested the money; showing wonderful cleverness and strength and perseverance. She has been the only stay of a large family, always ready to throw herself in the breach and pay anything that was needed. After buying the place in, she planted successfully for one or two years. Then the freshets began, and, after two or three very disastrous years of loss, she showed her good judgment by giving up planting altogether, and all that splendid rice land, under the finest, heaviest banks, is just returning to its original condition of swamp, growing up in cypress. She has land also covered with splendid timber, which must eventually be of great value, but as yet the money value of such things has not reached us, and the little shop continues to support a large family in their beautiful historic home, where with lovely flowers and beautiful oaks, every fence and hedge covered at this season with the glowing, sweet-smelling yellow jessamine, she leads a useful, contented, beautiful life, a blessing to all around.

I mentioned in my last letter that I had lost my good Jim, who had been with me fifteen years. I tried in vain to fill his place, but there was no one to be had that was reliable; so I got a mountaineer in August, paying his way down from the Blue Ridge. He promised well, but on the fifth day he was seized with nostalgia, and I had to drive him the eighteen miles to the railroad and put him on a train to return to his beloved mountains.

I would have had to return the eighteen miles alone on the road had I not met Jim, who was as pleased to see me as I was to see him; for the town life which his wife so loves is odious to Jim, and he asked permission to return to me until I got some one.

Being a person not easily daunted, I again engaged a mountaineer, not finding it possible to get a good darky, paid his way down and had Jim show him all his duties, the roads, etc., and he seemed a very hopeful person; and Jim returned to the hated town, satisfied as to my having competent help. Mountaineer number two showed no trace of homesickness for four months; but then suddenly, one day, it took him. It was no surprise to me. I knew it would come, and had got a black boy as help in case of emergency; so that when the attack came, I had Jake get the wagon and drive us to town, and I put mountaineer number two on the train. Then I increased Jake's wages and put him in charge of the stable.

Last week when I went to Casa Bianca to pay off, I took a niece and her two children, who were staying with me, and we had a very pleasant picnic dinner. The four-year-old children had never been in the country, and enjoyed everything, especially the lambs. Jake's home is about 300 yards from Casa Bianca avenue, and at 12 o'clock I told him that he could go and see his mother for two hours, and that I wanted to leave at 3:30 o'clock. Jake, however, did not return till 5:30 o'clock, though I sent after him; and, in his hurry, instead of calling the horses, which had been turned out on the lawn to enjoy the beautiful pasture, he ran them. Some one had left the gate open, and they dashed through it and never stopped running till they reached the gate at Cherokee, eight miles away, leaving me with my party of city friends, the sun setting and no horses to take us home!

Two men on the place owned horses, but they were turned out and could not be got up for some time. Besides, one was a terrible kicker and the other a runaway. I had to act quickly.

I said to the old watchman: "Go and tell the man with the fastest ox-team on the place to come here with his cart at once." In a very short time Nat appeared with a large black ox in a little cart of wonderful construction. I did not see how it was possible for a lady, two children, and a maid to get into it; but, apparently, it was the best that could be done. They had walked on, and I told Nat to go as fast as possible and pick up as many of the party as he could carry, and I would follow, as soon as the kicking horse could be put into a buggy, and take the rest. He assured me his cart could carry all, and went off at a rapid trot. After what seemed an age, Marcus came with his kicker, and with the wraps, lunch basket, and other encumbrances I got in and drove rapidly after the party, which was the funniest looking in the world; Nat running alongside and flourishing an immense cowhide lash, A. and the maid seated on a board which was balanced on the sides of the little structure so as to make a seat, the little boy sitting behind with his feet dangling and the little girl tightly clasped in her mother's arms. They had gone four miles in this wonderful fashion. As soon as we caught them, I made A. get out and take the children in the buggy, while I climbed into the ox jumper with the maid and told Marcus to drive home as quickly as possible, as the children should not have been out so late. I had been utterly wretched till I came up with them and found them all unharmed. Then my spirits rose and bubbled over. It struck me that the others were a little quiet, but I never knew the reason until we were all safely enjoying our evening meal. The maid was supposed to be driving, while A. held the little girl and Nat goaded on the ox, and at a very rough bridge the ox stumbled, the maid fell out, and the wheel ran over her, leaving the rest of the party without any hold on the big black ox! A most tragic situation, and such a mercy no one was hurt. It was very good of them not to tell me until afterward, and it was truly magnanimous of the maid to remark to Nat as she extricated herself from the ingenious contrivance, which he had constructed himself: "It surely is the handiest little vehicle I ever did see."

Saturday. I suddenly awoke to the fact after breakfast this morning that I had a note which was due at the bank to-day. It was pouring, and if I sent a check by mail, it would not be received until Monday, as the mail gets in after business hours. I called Dab and asked him if he thought he could walk down to Gregory and take an important letter to the bank before 2 o'clock. He answered promptly that he could. I got the letter ready and told him he could spend the night with his sisters and return by 1 o'clock to-morrow. He started armed with a large package of lunch and with my best umbrella and a dollar to spend. While I was taking my tea at 6 o'clock, to my surprise Dab walked in with a letter. He said Mr. S— gave him the answer and he thought he had better not go anywhere with that important letter, and so he had come straight back home! I was so pleased and cheered by this evidence of his sense of responsibility and fidelity to a trust. I had felt ill and miserable all day. I told him how pleased I was and thanked him heartily and told Chloe to give him a very fine supper after his walk of twenty-eight miles.

A rice field "flowed."

Sunday. I went to church this morning feeling very down,

which was wicked, for God's goodness is always there. When I looked around our little church, where a literal Scriptural quorum of two or three was gathered together, my eye was gladdened by the sight of a charming new suit of reseda cloth with a heliotrope toque! Then across the aisle I saw a cinnamon brown suit with a hat to match! Positively my spirits rose at once.

We are so accustomed to our mourning-clad congregation, nearly every one of us wearing black, we all know each other's very respectable costumes from year to year and watch with interest the successful and often ingenious remodelling of sleeves — I being the only recalcitrant who will not cut over sleeves, feeling sure that they will come back into vogue (which they always do before the faithful garment is laid to rest) — we never expect anything so astonishing as a brand-new tailor-made suit, and in colors too, and now to have the eye refreshed by two, is cause for rejoicing.

On Monday, April 18, I planted the wages field at Cherokee. Here we cannot so well use the machines, so I have the field sown by hand. I am planting mill-threshed rice in this field, which is an experiment on my part. In the autumn a buyer for a large rice mill in North Carolina came to make an offer for my rice; and he spoke of the "superstition," as he called it, of planters in this state that only hand-whipped rice could be planted to make good crops. He said the large crops made in Texas and Louisiana, which are practically ruining the rice industry in this section by keeping down the price, are the result of mill-threshed rice — none other is known or thought of. This made a great impression on me, for the whipping by hand is a very expensive process, more so than the actual cost of the work, because it gives such unlimited opportunity for stealing.

I had the habit formerly of planting twenty-five acres and dividing the rice; twelve and a half acres I sent to the threshing mill in a lighter, the other twelve and a half I had taken into the barnyard, stacked, and when thoroughly cured, had it whipped out for seed. The half sent to mill always turned out from twenty-five to thirty-five bushels to the acre; the part saved for seed turned out from fifteen to twenty bushels to the acre.

That happened several years in succession. I never have had a field hand-whipped turn out over twenty bushels to the acre, and I have seldom had one threshed in the mill until these last very bad years turn out under thirty.

All of this made me determine to try planting mill-threshed rice this year. I planted a small portion in a bowl of water on cotton, which is the approved way of trying seed, and nearly every single seed germinated and shot up a fine healthy leaf. So I felt no hesitation about it; and I began with my wages field, putting half a bushel more to the acre in case there should be some grains cracked in the mill. I went over early to the field and sat on the bank all day, while Bonaparte and Abram followed the sowers.

The women are very graceful as they sow the rice with a waving movement of the hands, at the same time bending low so that the wind may not scatter the grain; and a good sower gets it all straight in the furrow. Their skirts are tied up around their hips in a very picturesque style, and as they walk they swing in a wonderful way. This peculiar arrangement allows room for one or two narrow sacks (under the skirt), which can hold a peck of rice, and some of the sowers, if weighed on the homeward trip, would be found to have gained many pounds. They are all very gentle and considerate in their manner to-day, for a great sorrow has fallen on the family. Their tender, sympathetic manner is more to me than many bushels of rice, and I turn my back when they are dipping it out.

I have offered hand-whipped rice for sale at $1.30 a bushel, and mill-threshed at $1 per bushel, and have sold 159 bushels of the former and 225 bushels of the latter, which has been a great help. We have made a fine start on the upland crop, and the corn looks very well. The small acreage planted in cotton also looks well, and I hope it will be worked properly while I am gone.

May 9.

Left Cherokee for a month's absence, and drove to Gregory to take the through train to Washington, where I arrived the next morning in time for breakfast. I have a duty which calls me away. It was a pity to have to leave now, for the people had just become roused to an interest in preparing the land for their crop, and it is the first propitious season we have had for three years with no spring freshet, and I hope to get about 100 acres planted at Cherokee. I feel better satisfied to leave since Jim has returned to work with me and will take entire charge of the upland crop. His health suffered in the confinement of the town work. He was in bed a good deal of the time, and, what with lost time and doctor's bills, his wife found they were worse off instead of better, and finally, after nine months, she begged him to come and ask me to take him back, which I gladly did, and he has gone to work with enthusiasm.

While away, I visited Washington, Mount Vernon, Baltimore, and New York, and was much impressed by the immense strides made in every way since my last visit. The increase of wealth and luxury, the fact that simplicity of life is becoming impossible even to those who would prefer it, the rush and the hurry which one cannot avoid, the tyranny of fashion which no one seems able to shake off — all of these things amazed me. My good black Chloe once surprised me by saying: "You know, Miss Patience, ef yu aint een de fashi'n yu may's well be dead!" But Chloe follows at such a very respectful distance that the "fashi'n" so vital to her at this moment is a watered form of what was worn in New York four years ago. Still, I recognize in it the same note which I find dominant wherever I go and which is to me incomprehensible — it doesn't seem to me very self-respecting to feel obliged to follow some one else's taste so absolutely. One's eye naturally turns toward the changes of mode which are pretty, but to feel bound to follow simply because fashion decrees, I do not understand.

I saw many things that interested me greatly. One evening I was walking back to the St. Denis about 10:30 when my escort said: "That scarcely seems possible at this season." "What?" I asked. He pointed to a closely pressed row of men in a single file, on the edge of the pavement, one immediately behind the other in perfect order: decently dressed, respectable-looking men. It had a strange look to me, and I asked the meaning of it. "That's the Fleischman line." This conveyed nothing to me. "It is a great bakery here, which for years has distributed every night at twelve all the bread left over from the day's bake, one loaf to each man. I know that in winter the line extends many blocks, but at this season I am surprised to see such a line at this hour; it will be twice as long by midnight." My heart just stood still as I looked at it.

That so many men, looking so respectable, could need a loaf of bread, and wait silently, patiently for hours together seemed impossible to me. Where I live there is no hunger, no want; life is so easy, food so plentiful. A few hours' work daily feeds a man and his family.

One day Jim was driving to town to spend Sunday with his family, and the next day he told me that he had met an old woman on the road going from one plantation to another. She seemed half blind and looked so miserable that he stopped and asked her where she was going, and offered to take her there in the wagon, as he had to pass right by. He helped her in and she told him she was very hungry — had eaten nothing since the day before.

"Oh," I said, "Jim, did you give her something to eat?"

"I didn't have nothing to eat with me, ma'am, but the sticks of candy you giv me to take to my chillun; but I giv her them, en you never see any one so please'." Then he went on to say: —

"It seems to me sence I ken remember this is the first person I ever seen real hungry."

"You mean you have never met a hungry person on the road before?"

"I never met none on the road nor never seen none nowhere that was perishin' with hunger."

I was scarcely surprised, for my mother always at Christmas told her man servant to find out the poor and those needing food that she might supply it. The old man always reported that there was no one he could find in need of food, but plenty to whom a present of tea and coffee would be most acceptable, and to these the packages of sugar, coffee, tea, and tobacco always went.

The next thing I wish to mention is my visit to the Agricultural Department in Washington. I went in search of information as to the planting of alfalfa, and the use of the impregnated soil, which is said almost to insure success. That is the crop to which I look with much hope for our uplands, and I have much at heart to take in a beautifully drained area of thirty acres which has been pastured for some years and plant it eventually all in alfalfa. At first I could not get more than ten acres in fine enough condition, I suppose; but all that I can plant this year I wish to. I have already bought the wire to enclose it, which is a heavy outlay, and had the cedar posts got out, so that it will not cost much to get the fence put up; but I have no proper disk harrows and cultivators to put the soil in the best condition, and the outlay is too heavy to venture on buying them, so I will do the best I can with my old-fashioned implements and plant a heavy crop of cow-peas on the land as a preparation for the alfalfa, which I will not plant till September.

It was a great pleasure and satisfaction to find men of intelligence and education whose whole time is devoted to the effort to promote the productiveness of soil everywhere, and to find them willing, I may say eager, to assist me in every way. All the information was given in a brief and yet courteous way that was a great boon to me, and the reading matter furnished me by them on the subject will make it plain sailing, if only I succeed in getting good seed; and the impregnated soil, I believe, will prove a blessing to this section and solve many problems.

On my return after a very hot journey I reached Gregory at 10 o'clock at night and drove to a pineland two miles away, where I was most hospitably received and spent a delightfully cool night. The heat in Washington and New York had been extraordinary for the season. The next morning I attended to my business in Gregory and started on my homeward drive of twenty miles about 10 o'clock. I drove first to Casa Bianca, where the June rice wages field of twenty-six acres was being planted. I found Marcus and the hands in fine spirits. The April rice was very fine, they said, especially the River Wragg, though Marcus told me it was suffering greatly from the need of hoeing, but he could not stop the preparation of the land for the June planting to hoe it out. This trouble is due to the moving of so many of the young people last winter to town. They were all good hoe hands and there is no one to take their place. The men now think it beneath them to handle a hoe; that they consider a purely feminine implement; the plough alone is man's tool.

I stayed at Casa Bianca until 3 and then drove to Cherokee, where everything had a very different aspect. When I drove into the barnyard, after the usual exchange of politeness with Bonaparte as to the health of each member of the family, I asked him how the rice crop looked. He laughed in a scornful way and said: "W'y, ma'am, I may's well say der ain't none." "No crop, Bonaparte; what do you mean?" He continued to smile in his superior way, and the hands standing round chimed in: "Yu' right, Uncle Bonaparte, you may's well say dey's none, we fiel' ain' got none tall een um, 'tis dat mill trash rice; you kin see de rice dead een de row wid de long sprout on um, all dead." I answered quickly, "If the seed is dead with a long sprout on it, that proves conclusively that the seed was not to blame; if the seed had been defective, it would not have sprouted. There was a good stand in Varunreen; before I left, the sprout water had been drawn." "Well, ma'am, dey ain't nun dey now to speak of." "How is the tide now?" "Most high water now, ma'am." "Get my boat out at once and I will go over and see for myself."

The hoe they consider purely a feminine implement.

While they were getting the boat one of the hands asked me to give him seed to replant his land. "What is the use if you think the seed is bad?" "We want you fer buy mo' seed. Col. Naples got seed fu' a dollar en forty cent a bushel." I told him I had no money to buy more seed; I was willing to give them more of the same rice which I had if they wished to replant, but that was all I could do.

I got into my little white canoe, which I call the "Whiting," and had Bill, one of the most pessimistic renters, to row me down the river. The tide was high. I was able to step out on the bank without any sticking in the mud, which makes it such a horrid trip when the tide is low. I was greatly relieved at sight of the field. There was plenty of rice in it, as I pointed out to Bill as he walked round the bank behind me. The rice was stunted and growing poorly, and upon inquiry I found that it had been dry during the month I was away. Though not a drop of rain had fallen in that time, it had not been moistened by letting the water into the ditches, as it should have been from time to time. That would have made all the difference in the growth of the rice; but foreman, trunk minder, and hands were all so sure it could never make a crop, being mill-threshed seed, that they have not given it a chance, content to declare loudly that there is no rice in the field.

I am greatly comforted by the sight of it, for there is plenty of rice there to make thirty bushels to the acre should no disaster come to it; and I get into the little "Whiting" with a quieter mind, though still greatly distressed about the hands' rice. The row back is most refreshing, there is such a breeze, but the sun having gone down suddenly, the damp chills me, for I had not thought of taking a wrap, it was so hot when I left the wagon. I give orders to Bonaparte to have the field hoed out at once so that the water can be put on as soon as possible. Then I interview the trunk minder, whose business it is to water the rice, and ask the meaning of this talk of there being no rice in the rented fields. He begins about mill-threshed seed, but I show him the glass dish of rice in which every grain had sprouted and grown vigorously. The sight of this seems to confuse him. Then he mentions that he had got a bushel from me "to plant out to his house een a bottom," and that he never saw a prettier show than that patch of rice. "Then," I say, "you see it is not the seed; you must have left that rice exposed in some way to the hot sun just as it sprouted." "Dat's a truth, my missis; it must be so. I did shift the water and I must ha' left it off too long, an' the sun took effec' on de rice w'en 'e was sproutin'."

The result, however, is the same. In the three fields of rented rice the stand is so poor, they tell me, as scarcely to warrant cultivation further. The hands, to begin with, I am told, carried home to eat much more than half of the rice given them to plant. They always take home a goodly portion on the principle that a bird in the hand is worth two in the bush; but on this occasion, as it was mill-threshed rice and was not coming anyway, and I was safely away in Washington, they scarcely put any in the ground.

Thoroughly disheartened, I got into the wagon and drove to Peaceville, the little pineland settlement, just as the night fell. The dogs give me a joyful, noisy welcome and Chloe seems overjoyed to see me, while little Imp shows every white tooth in his head and his black face beams with joy. Chloe has a delicious supper for me, to which I do full justice, not having eaten anything since breakfast, at 6:30. The bungalow is very comfortable, though not much for beauty — the servants have moved all my belongings from the plantation while I was away, and I find everything I need except my piano and my books. The piano could not be moved because Jim has had the team in the plough every day. They have done very well, however, for the piazza is filled with blooming plants, and the house looks clean and cool in its fresh white wash. The pineland is noted for its pleasant nights, and I woke refreshed in the morning, but to find I had taken a terrible cold in my homeward progress on the river, I suppose.

Wednesday, June 15.

I drove down to Casa Bianca to-day to see how the rice looked and to give orders for the bringing up of mutton weekly. I have been so entreated to furnish the village with mutton weekly again this summer that I have consented to do it, though it is quite an undertaking to have it brought up the twelve miles regularly and early enough in the morning. Marcus met me with a very solemn face, and when I said in my cheeriest voice, "How is everything, Marcus?" he took off his hat, made a low bow, and said:—

"Miss, I have very bad news to-day."

"Oh, Marcus, what has happened? Is Rubin dead?" Rubin is a very beautiful bull, the pride of the place. Very slowly and with great dramatic effect Marcus answered: "No, ma'am, but the crop is ruin', all the rice is gone!" "Impossible," I cried, "it was so fine the last time I was here."

"Yes, ma'am, but Sunday dey come a sea tide, what just sweep over the bank an' 'e bin on de rice till now; de watah bin a foot deep on all de rice, an' salt, ma'am, salt like 'e'll do for cook with, en to-day fu' de first 'e begin for drop, en I giv' yu' my word, ma'am, in my fiel' de rice yu' kin see on de hill is red, same as red flannel! You kin come en see fu yuself, ma'am, down as far as de bridge, for you kyant walk on de bank, 'e too wet."

I went and saw that the entire place was flooded and that the hills as they peeped out, here and there, had a reddish hue, instead of the vivid green of healthy rice.

What a disaster! A bolt out of a clear sky. If Marcus is right, it means ruin, and up to this time the rice was splendid. Of course, if salt water has covered the rice since Sunday there can be no hope except for the June rice, 60 acres, which was still under the sprout water, which the sea tide only diluted, and so it may escape. Marcus was so thoroughly cast down that I had to cheer him, and searched my brain for grains of comfort for him, until by dint of effort I became quite cheerful myself, in spite of the very black outlook. I made him taste the river water and he reported it still salt.

I stopped at Cherokee on the way home and saw the corn

and cotton, and they are beautiful and do Jim great credit, for it is only by the constant stirring of the land with plough and cultivator that the crop has not suffered from the six-weeks drought. The oats are being cut, much injured by the drought, having made no growth and the grain not having filled out.

Tuesday, June 21.

Gave orders yesterday for the threshing of the oats to-day. The engineer fired up at daylight and had a fine head of steam on when the hands assembled, but Bonaparte looked at the sky and said he thought the day was too "treatish to trash" and sent the hands away, at least a quarter cord of light wood having been wasted, besides the engineer's time and the waste of the hands' time, and, worst of all, the losing of the day when there is so much work needed. It did not rain at all, and even if it had rained there would have been no harm done, for I had purposely had the oats hauled into the mill the day before, so that in case of rain they could still thresh. When I drove down, expecting to find things in full blast, I was very much provoked. I just had to leave, for there is no use to give vent to one's wrath. I told them to thresh to-morrow without regarding the weather.

June 22.

Went in to find threshing successfully accomplished; they got through quite early, so I determined to let them finish out their day by moving the piano. The oats made twenty bushels to the acre, which is more than I thought possible. No one in a city has any idea what the moving of a piano is. I always feel as though I were personally lifting and handling it, so entirely is the responsibility on my shoulders. My upright piano is my most cherished possession, companion, and friend, and I am always nervous over the perils of its four-mile drive from plantation to summer house.

A small mattress is put in the plantation wagon — I have

no spring wagon — and on that the piano is put and steadied by two men while it is slowly driven out. It always takes eight men, as they are not accustomed to lifting, and they make a great ado over it. Just as the piano was lifted out of the wagon up the rather high steps on to the piazza at the pineland, they set it down at the head of the steps, and gave it a great push to roll it toward the sitting room door, there came a tremendous crash. The piazza had fallen in on the side toward the house. Fortunately there were no men in front of the piano; they were all behind. I was

The back steps to the pineland house.

standing very near and called to them to hold on to it a moment. I had two heavy planks brought and put as a bridge from the place where the piano rested into the door, and as soon as they got the front rollers on the plank all danger was over, but for a time it looked as if there must be a terrible smashup. I sent one of the men under the house to see what had caused the crash. He reported it was the giving way of one of the blocks, which was so rotten it had crumbled away. "Why, Bonaparte, I sent you to examine the foundation of this house and see if any repairs were needed, and you said it was all in order." Bonaparte only murmured apologetically that he was too busy to see about such small matters.

I am very thankful no one was hurt, and the dear piano is safely installed. It is a small Steinway upright and is very nearly human in its companionship. This is the sixth Steinway I have had, I believe, having never owned any other. I watch its health with desperate anxiety, for it will have to last me to the end, unless something wonderful happens to revive rice. I have been at the piano all evening and it is now 1 o'clock and I am too much excited to go to bed, and that is why I am writing.

An unknown friend sent me two years ago two volumes of Russian music that I find fascinating. A cradle song by Karganoff and a prelude by Rachmaninoff especially possess me. When I play that Berceuse I feel myself the Russian peasant clasping her child, with intense strained nerves, always alert, in spite of the soothing, delicious melody she sings and the reassuring loving reiterations of promised safety. The prelude is tremendous, foreshadowing awful depths of pain, endless struggle through distress and discord, up, up, creeping to a final chord of perfect harmony, but a minor chord. If I were asked for what I was most thankful in my possessions I should say my power of enjoyment. Here entirely alone, with never any audience — and in some measure because of that — these things can fill me with such intense pleasure that it is like being on a mountain top with the heavens opening in a glorious sunset, revealing to the panting soul the inner Court of the Beyond.

I remember when taking a singing lesson as a girl being so

overcome by my inability to express what the music said to me that I broke down and was reduced to tears and said: "Oh, Mr. Toiriani, there is no use for me to go on: I have no voice and it is useless." He turned fiercely upon me and said: "Voice — what does that matter? You must go on. Vous avez le feu sacré." As I used my handkerchief violently in my effort to suppress the sobs that would come, it seemed to me a poor consolation, for if the said fire found no outlet it must consume and not illuminate; but I dared not answer, only struggled for composure to go on with "Buona notte, buon dormir" in a feeble, quavering, high soprano. But I often think now I understand more what he meant. One is independent of outside things; there is a warmth and a glow and a depth that fills and satisfies, irrespective of results and externals.

July 3.

I paid off this afternoon, as the Fourth of July is the day of all days the negroes celebrate. It was always so before the war. Every creature has to be finely dressed.

Chloe came in yesterday in great excitement to say Miss Penelope had opened a big box of the most beautiful hats and she wanted the money to buy one, "Quarter of a dollar and 10 cents." I exclaimed at the cheapness, but when she returned and showed me a very large, black straw trimmed with a wealth of black and white veiling and a huge purple orchid on top I was still more filled with wonder how it is possible.

Chloe is perfectly happy. The cloud which has hung over her for the last week is dispelled by the consciousness that she is suitably provided to celebrate the country's birthday.

July 6.

On Sunday sent word to all the Cherokee hands that I wanted them to hoe the rice yesterday. Of course no work is ever done on the Fourth. It is a day of general jubilation among the darkeys — gorgeous costumes, little tables set about with ice cream, lemonade, cakes; every kind of thing

"A very large black hat."

for sale — watermelon above all. Yesterday morning there were three women in the field and a boy to hoe the rice. The other hands sent word that "they couldn't work so soon arter the Fourth."

I am anxious to get the field hoed out, for besides its being very grassy, I see by the weather report that the river has risen to 18 feet at Cheraw and is still rising, and it is most important to get the rice clean of grass and the water put over it before that freshet water gets down here. There are 12 tenant houses on the place with 20 grown hands and about 6 half hands and numbers of children. These people, by their agreement, are to work for me whenever I call them. When I do not need them they are at liberty to work wherever they choose, but when I call them they are bound to come. With this understanding they have their house free of rent with an acre of rich upland, all the wood they need, winter and summer; yet it is impossible to get the necessary work done. One-half of an acre is the task for a whole hand in hoeing rice, and the eleven-acre field should not take more than two days at the utmost; but at this rate it will take more than a week, and I am powerless to command the work. I pay in money always, but they prefer to make excuses of illness to me, and slip off and work for my neighbor who pays in cards redeemable only in his store, because it is done on the sly and in opposition to my authority — in other words, that is freedom.

Who could succeed with such a state of things in any business? I am so discouraged. I do not see where it is to end. This fertile soil must just grow up in weeds and go to waste, because, forsooth, life is too easy here. Why work when one can live without? Why carry out a contract when one can wriggle out of it as a snake does from the effort to hold him with a forked stick, and go and bask in the sun and satisfy the elemental needs as the snake does?

Jim came in to-night, having done a fine day's ploughing, but very angry because he had spent his midday hours chasing three negro pigs through the corn-field. He says they are in the field every day doing great damage, and he cannot find any hole in the fence where they could get in. My own twenty-five pigs are kept confined in a crawl or pen for fear of their getting into the corn, and these robber pigs are fattening on it. I very much fear some one in an unseen moment turns them in the gate, for they belong to people living on the place.

I am broadcasting peas on the field from which the oats were cut, and these pigs are eating them before they can be ploughed under. I cannot bear to order the pigs shot, but suppose I will have to do it. Things have never been so bad as this before. There is some influence under the surface of which I know nothing. God only knows how it is to end, and it is a great comfort to feel that He does know everything and never fails them that trust Him.

July 16.

Our rector came to us for service to-day, and we had an excellent sermon from him. To-night all the neighbors assembled at my house, as usual on Sunday evenings, for sacred music. It always moves me to see the delight every one takes in this very simple way of passing the evening — men, women, and children are equally enthusiastic. I tried having the children in the afternoon, but they did not enjoy it as much, so they all come at 8 and sing until 10, and then home to their night's rest.

July 17.

I was scarcely able to get back from Cherokee to-day, so exhausted was I. Ever since I came home I have been rising at 6 and going into the plantation very early, attending to my work, and getting back to the pineland between 12 and 1 — a rag in every way. I am so much the worse for wear that I shall have to give up for a while; as most of the important work is over it will not matter so much. Have had great trouble about the wood again — but will not go over it.

July 18.

Too unwell to go out, so have spent the day in the hammock, reading with great delight Mrs. Gillespie's "Book of Remembrance." What a great thing for a woman like that to leave such a record behind her! How I wish all the great women who have passed out of our sight and hearing could have done the same; it would be an inspiration and help to the poor things growing up, with their confused ideals.

July 19.

As I was still too weak to go out this morning, and the mercury was 94 degrees, I sent Chloe in to the plantation, driven by the Imp. With the big umbrella over her she was most comfortably arranged, and I told her not to hurry back, as Gerty could cook what little I needed. For some reason the trip went against her, and she came back in a very bad frame of mind. She said some one had jeered her on the road, and said she had given up the job of cooking to Gerty, and that I must want to kill her to send her out in such heat.

I really am mystified, for I have gone in every day since June 7, and many days have been hotter than this. Last summer very often Chloe walked in to the plantation and saw after things in the garden, and walked back, in spite of remonstrance on my part. I wish I knew who had jeered her, but I really cannot bring myself to ask. Better just to let her alone until she recovers her equanimity, but it is trying when I am feeling so below par. I certainly shall not send her again.

July 30.

Rose at 6 and went into the plantation soon after 7, and got through my work very comfortably, though the thermometer was 96 degrees when I got back.

Am reading Sir Walter Besant's autobiography with interest. While I was in Washington some one gave me a list of books from the Philadelphia bookstore, and the reduced prices have made it a blessing. I have been able to get books otherwise entirely beyond my reach, and it is such a treat.

I think biography is the most fascinating and satisfying reading. I sent on $7, and no one could imagine what a number of delightful books came back. I positively gloat over them. I have not had such riches in years.

We have a very nice little book club in Peaceville, established about eight years ago by the thoughtful kindness of a friend, who had visited me several times and became greatly interested in Peaceville and its "old time" atmosphere. She sent all the novels her family had finished reading. And her sister-in-law, who lives at the North, but was making a visit South and was there when the books were packed and sent, subscribed to McClure's for the Peaceville Book Club, and has kept it up ever since; also, from time to time, sending a well chosen new book. It is so very kind.

When the first lot of books arrived I went around, mentioning that this donation had been made. I said I hoped every one in the village would join, and that the membership would be 10 cents. Every one was roused and delighted, and there was much discussion as to where the books should be kept. Finally the postmistress, who occupied a little cottage where the whole village assembled to chat while the mail was being divided, consented to keep the books. This was an ideal arrangement, and I had a large bookcase, simply made, with lock and key, and the books were installed.

After a short time two of my dear friends, whom I had thought of as especially likely to enjoy the books, came to me and said they would not be able to join the book club. I wondered, and urged them to join, when the mother said: "We had intended to join, for we thought the fee was 10 cents a year; but since hearing it is ten cents a month it will be impossible for us to indulge ourselves in that pleasure."

"Oh," I said, "it is ten cents a year and not a month. The person who told you made a mistake — " Then they said, "That is too delightful; for we felt miserable at having to give it up."

Now the fee had been merely in order to give some little control over the books, and I had thought if it was a monthly fee we should be able to get some books every year with the fees; but as it was established just for these very cases, I suddenly changed the plan, and 10 cents a year it has remained ever since. The kind friends have continued to send the novels they have read, and some of their friends in New York have sent boxes of magazines, so that the little club now has about four hundred volumes. Our dearly beloved postmistress has gone to join the majority, and the little cottage is closed, and the books have been removed to a shed room in my house, but there is no estimating the pleasure they have given, and still give, and the weary hours they have relieved.

Every year I get one new book with the fees, and this year, thanks to the wonderful Philadelphia bookstore, I got three, for novels are preferred at the Book Club.

When people work hard and have little pleasure they need relaxation, which means "Mrs. Wiggs" and "Lovey Mary" and tales of chivalry and wonder and social joys, and all the things which every one longs to have for themselves and their children. The writer who can blot out all the sordid present and raise one into a different atmosphere and keep one there for two hours is a mighty benefactor.

August 6.

I have suffered so from heat that I felt distracted. Went to the plantation most reluctantly, but it was a relief, for it felt cooler driving, but poor Ruth suffered greatly. I am ashamed to be so knocked down by the heat; my mind seems addled.

Bonaparte's daughter-in-law, Kiz, is very ill with typhoid fever. Her husband brought her to the doctor to-day in an ox cart. The doctor was very angry, saying she was quite too ill and that he must take her quickly home.

I had made some jelly for her and sent Patty running after the ox cart with it. She said Kizzie was very grateful and took it all, saying it was the nicest thing she had ever tasted.

Her husband brought her in an ox cart.

Poor, poor soul; in this heat — no ice, no anything that she should have! I who am quite well miss ice terribly, and think of her with that fever!

I sent some jelly to old Amy, too. I do not think she can recover. She is Patty's grandmother. MacDuff feels the heat greatly. The mercury has been over 90 for several days. The colts both have distemper and cannot be driven for a long time.

CHAPTER III

September 3.

IT is time for my harvest to begin, but for some reason the rice is ripening very slowly, and I fear the first field at Casa Bianca will not be ready to cut before the 14th of this month. It has never quite recovered from the salt water and is not as fine as last year. At Cherokee one field of rice is very fine, the other not very good; but the corn is of the best, and so are the peas. A splendid crop. In July I took up thirty acres of very well-drained land, enclosed it with an American wire fence, and planted some of it in cow-peas preparatory to planting alfalfa this autumn. The peas are most luxuriant, a solid mass of green about two feet high. They show the benefit of the subsoiling I had done, for I used no fertilizer of any kind on the land. I have gone to great expense to put this land in good condition, for I have great hope of making alfalfa our money crop in the future — poor, dear rice seems to have resigned that position.

September 13.

Mr. and Mrs. S. from Indiana are staying with Mr. L. They came to look into the possibilities of this country for cattle raising, Mr. S. being one of the most successful and best-known breeders of Hereford cattle. He wishes to see as much of the plantations as he can, so I invited them to spend the day with me at Casa Bianca, as it is a good natural pasture. I took down everything with me for a nice luncheon, and they seemed to enjoy the day. Mr. S. said my cattle were in fine condition, and that the grass was very good.

While they amused themselves wandering about the grounds and over the rambling old house I went to see Marcus. He told me he had all the hands he could get minding birds and picking grass out of River Wragg and that he had taken the water off to-day as he hoped to cut it day after to-morrow. After lunch when we went out the look of everything had changed — it had been a perfect morning, with little white clouds flitting about, just making you wonder at the blue of the sky in contrast to their airy whiteness, but now they had heavy dark edges and they rushed heavily and wildly about, and there was something in the air that made one sniff a coming storm. Mr. L., who knew the signs well, asked me to have his carriage got, and left at once, advising me to do the same; but I had some things to attend to before leaving, and so was nearly an hour late. I told Marcus to put the water back on the rice or it would be whipped to pieces by the wind, which was now tremendous.

My twelve-mile drive home in an open wagon was a race with the storm, wildly exciting and exhilarating, in spite of the danger from falling limbs and flying branches. All along the way the cattle were gathered in the middle of the road, and my companion said she had always heard that was a sure sign of an approaching storm; ordinarily they are in the woods and I was greatly surprised at the number. I knew the negroes owned a good deal of cattle, but did not know there were such herds.

The horses were greatly excited and it did not take us long to reach home. Though it had rained all the way it did not pour, and the wind being so high seemed to blow the rain away, and we were very little wet.

The wind increased in violence every hour, and now at 10 o'clock it is a terrific gale. I have been all over the house examining windows and doors to see that the fastenings are secure, and am going to bed, for I am very tired.

September 14.

The storm raged terribly all night; sleep was impossible. The rafters creaked and groaned, the windows rattled, the house shook, the wind roared through the pine trees, while the cracking of limbs sounded like musketry and now and then the loud thud of a falling tree like cannon. These sounds kept the ear and mind on a prolonged strain. In the dawn of the morning I looked out — a gloomy, dark sky, trees down in every direction, not a fence in sight; but no houses down.

Later in the day I went forth to find out how my neighbors had fared, and found every one so thankful to find themselves and their families alive and unhurt that every one was cheerful and bright. Most people sat up all night and all seemed to have had me much on their minds.

"Such a terrible night for you to be alone in the house; we thought of you constantly." I had been thinking with such anxiety about people on the islands and at sea that I did not feel frightened for myself; but I found the servants had been very anxious about me, and Jim had walked round the house several times, but finding all still and no light had gone back to the servants' hall. I hear of many marvellous escapes, houses falling and pinning people down, without a single death and with little injury.

All the planters went out very early to the plantations, carrying axes to cut their way along.

September 15.

I rode to the plantation to-day, as the road is impassable for a vehicle from the village to Cherokee. There the storm played havoc; the immense oak trees are down in every direction; some uprooted, some split into several sections. Just back of the dwelling-house there was a large oak heavily draped with ivy that had been snapped off in falling, narrowly escaping the house. Two other very large oaks to the north-east of the house are down; they evidently broke the force of the wind on the house, which is not injured at all. The earth is strewn with gray moss and small green twigs and leaves, so that it looks like a huge gray and green carpet.

A two-story barn is down at the barn-yard, also another building, and the screw is badly twisted and in a falling condition. The corn, which was so fine, has been torn and tangled, and a great deal is lying on the ground partially buried in the mud, the heavy beating rain having left the fields almost boggy. I sent all hands to gather up the fallen ears; in the barn I had them shucked and spread over the floor to dry. At the point of nearly every ear the corn is sprouting. Of the cow-peas about ten acres are ruined; they were loaded with pods almost ready to pick, and they have been stripped of leaves and fruit and are only bare stems. I have never seen a storm so thorough in its work and so minute in its attention to detail.

I always try to see the grain of comfort in every misfortune, and find it now in the thought of the profit to the land in that heavy mulch of pea-vine leaves and pods; meantime I will not make seed. Fortunately my alfalfa peas were younger and in a sheltered situation, and have not been at all hurt. I had not heard from Casa Bianca until to-night, and the same tale of destruction and desolation comes from there, but there has been no loss of life. I was so afraid some of the negro houses might have fallen and hurt some one, for they are very old. There is not a fence standing, and the demand for nails is great.

September 17.

This morning old Maum Mary came to bring me a present of sweet potatoes, indicating that she was in great need of nails; so I made her a present of some nails and also a piece of money. She and old Tom live on a little farm of their own, where they plant a field of corn, a patch of rice, a patch of cotton, and one of tobacco. They raise three or four hogs every year and have a cow. In addition to these they have a most prolific pear tree and a very large scuppernong grape-

"Old Maum Mary came to bring me a present of sweet potatoes."

vine, and the sale of their fruit brings them in a nice little income.

After the interchange of presents had been made and she had eaten the plate of meat and bread and drained the cup of coffee which I brought her, her tongue was loosed, and she said: —

"Yes, my missus, I neber see sech a judgment on de tree! De big pine 'ood is lebel down, en I had to climb for get yuh, but I ain't hab a nail, en de fench bin down, en me tetta, en me little crap o' corn bin dey open, en ef de Lawd didn't bin dat mussiful dat night en confuse de critter mind, all 'ood a gone. Yes, my missus, eberybody fench bin down, but not a cow nor a hog ain't eat nothing. Ain't yer see? De Lawd confuse dem mind to dat; Him is mussiful fer true. Dat night, my missus, de house shake en rock so, tell me en Tom git up en set down by de fiah, en we pray, en we pray, but de fiah cu'dn't burn, kase de rain po' down de chimbly. We de pray, en de house de rock en de shingle de fly, bam dis way, en bam dat way, en Tom cry out en 'e say, 'Yes, my Lawd, we is sinna fo' tru, but spare we dis time,' en den I teck up de disco'se en I say, 'Lawd, I know I is wicked, but gi' me anoder chanst.' En de wahter gone through de house, en de shed blow off, en de wedder-boa'd blow off, en de tree all round de crack en de fall, en, my missus, w'en de mawning come I was susprise w'en I see Tom de day, en me de day, en de house de day, en I hol' up my han' en I cry, 'My Lawd, yer is too mussiful, yo' jes' trow down de boa'd en de shingle, now ef dat bin a man, strong like a you, him 'ood a throw down de hol' house.' Yes, ma'am, I'se tankful to de Lawd," and with a deep courtesy she went to mend her fence.

September 20.

Harvest is going on at Casa Bianca — the much-tried River Wragg field is being loaded into the flats, in spite of its being soaked in salt water in July and swept by the gale last Tuesday. I cannot help hoping it may make something, it looks so pretty and golden as it is being "toted" into the flat. The night of the gale I thought it would be completely destroyed, because it was dry, but a tremendous tide rushed over the banks and topped the rice, thereby saving it from complete destruction. The June rice, however, fifty acres of which was very fine, has been greatly injured by the topping tide, for it seems the water was brackish, and the rice was just in milk. Marcus was bragging about this rice, and my hopes were high, but now he shakes his head and looks solemn.

Some years ago a lady in Saratoga said to me: "The Lord does not seem to have much respect for you rice planters." I answered: "I think Job's friends and acquaintances said the same thing to him."

Certainly it behooves us to imitate that worthy's patient endurance of the calamities which fell so thick upon him for years, and his firm faith in his Maker.

In the Old Testament the promise of worldly prosperity as a reward of obedience to God's law was very distinct, but in the New Testament it is different — sorrow, adversity, tribulation, are mentioned, and the promise is of peace within, of power to be undismayed by seeming disaster, strong in the faith that He doeth all things well.

> God moves in a mysterious way
>   His wonders to perform,
> He plants his footsteps on the sea
>   And rides upon the storm.
> *   *   *   *   *   *   *
> Judge not the Lord by feeble sense,
>   But trust him for his grace;
> Behind a frowning providence
>   He hides a smiling face.

Several of my friends in the village are ill, and fresh milk is much needed; so I waited till after sunset, when Gibbie had finished milking, to take the fresh milk with me. It was so little that after sending out three little pitchers there was none left for myself. Gibbie is doing his best to dry up the cows; this was the last trial.

In the morning I found Eva had not come out to do the work I had pointed out to her, and I went out to the street, meaning to go to her house and see what was the matter. I found no gate to her large enclosure and could not get in, so went to Gibbie's house to ask the way. It was about 11 o'clock and Gibbie was supposed to be at work. Saw the children and asked for their mother, but they did not seem to understand, but when I repeated my question the little one answered: —

"Pa dey een 'e baid."

I looked through the door and there was Gibbie fast asleep across the bed. I went in and poked him with my parasol, but he did not wake, so I left the house feeling hopeless — how can any work be done with this going on!

As I went through his yard I met his wife carrying a burning coal between two sticks. She had been over to a neighbor's, as she said, "to ketch fiah fo' cook Gibbie bittle." She directed me to her mother-in-law's house through a labyrinth of fences and gates.

I was much interested, for it is just what Stanley describes in "Darkest Africa," a system of passages of stockades, making hasty entrance impossible and so guarding against surprise; any one finding his way through must be seen by the inmates before reaching the innermost barrier. I wound my way through a field of splendid potatoes, then through one of peas, then into a field of splendid corn with peas running to the top of the stalks loaded with pods seven and eight inches long.

I went into the house, where Nobby, Eva's youngest son, a youth of 18, was sitting contemplating a big sheet packed with peas which lay on the floor. I asked where his mother was. He said in the field behind the house. He re-

"Pa dey een 'e baid."

mained sitting while I went round the house, where grew luxuriant tomato plants loaded with fruit and very tall okra, and on to another fine potato patch, where there were also peas, which Eva was picking.

She was much startled at seeing me. When I asked her why she had not come to work as she promised she hesitated and stammered, then said that the cow broke her fence and she had to stay home to mend it.

"Surely that big idle boy Nobby could mend the fence," I said.

The fence showed no sign of damage, and I knew she had just preferred to stay at home. I spoke severely and told her to come to-morrow and do the work. She has in all about ten acres with her house, and her agreement is to give me one day's work every week as rent, and she cannot make up her mind to do that if she can possibly escape it.

My only consolation was the extreme abundance and comfort of everything and the cleanliness of the houses and the children, but that is a great comfort to me.

I have made myself a beautiful big blue denim apron turned up about twenty inches, so that when I go in the field to get rid of the cockspurs and see the work I need not be idle.

My field of pea-vine hay is beautiful, but it was so badly ploughed that here and there cockspurs were not turned under and they would ruin the whole field. I have paid a woman twice to go through the field and pull out the plants before the fatal little burr was hard. I went through it myself some time ago and found that she had only broken off the heads and left the roots, all there to spring again.

I pulled out quite a number, and to-day called Dab to go into the field with me to pull them. If only I had told him to bring a hoe the day would have been saved. In order to get to the field by the shortest way I had to pass through a low spot in the corn-field which was grown up with weeds dense and as tall as my head. The ox cart had made a track in the midst, where its wheels had mashed the weeds, from the barn-yard. I was about fifty feet in front of Dab, lifting my foot very high at each step and going very slowly, with eyes everywhere, when six feet in front of me I saw a heart curdling sight — a moccasin so enormous that I could not believe my eyes.

He lay with his tail a foot beyond the wheel tracks on one side and his awful head a foot beyond on the other! I called as softly as I could to Dab, who was just opening the gate, "Bring a strong stick quickly to kill this snake!"

Dab called aloud in his most educated tone, which he very seldom uses, "A snake, eh? What kind of a snake? A big snake, eh?"

"Come at once, Dab, with a strong stick!" I said in anything but a conversational tone, but Dab continued to discourse and ejaculate, and before I could get him to take a lath from an old gate near which he stood the monster, who had listened to everything, slowly moved into the thick bushes and was gone.

There I stood, afraid to move one way or the other. I do not remember ever to have been so thoroughly demoralized since I was a child. When Dab came up even the tail was out of sight. I hate to think it, but it almost seemed as if Dab had dallied and waited until he was sure it had gone, for I kept crying, "Come quickly, it is beginning to move! Oh, Dab, come on, it will get away! It is going!" and not until I cried in despair, "Now it is gone!" did he come forward with great boldness, a splendid lightwood stake in his hand with which the snake could easily have been killed while it was in sight. I would not let him pursue it into the high growth.

I sent him back to the house for a hoe, and while he was

gone I stood there battling with myself. I could not bear to go on through that tall, dense growth of grass and weeds with this terrible thing somewhere, but I said to myself: "You have never let fear turn you back from an undertaking in your latter life; are you going to turn craven now? If you do you will be miserable; your life is beset by many dangers; once let fear get the upper hand and your composure and peace of mind are gone."

So I argued and reasoned and fought with myself, and by the time Dab came, it was easy to go on. I took the hoe from him and cleaned a space of weeds in the direction the snake had taken, and when I had showed him that I was not afraid to do it and how I wished it done he took the hoe and very gingerly chopped down the growth toward the vegetable garden, for I feared very much that the monster should establish itself in there. I kept behind him, encouraging him on, when he gave a shriek and cried : —

"Der de snake now." No educated tone now. He cried aloud "de snake, de sing."

I tried my best to see the snake, but could not. He is a little taller than I am and could see over the bushes.

"You must kill it, Dab!" I said. "If you do not it may bite you some day when you go to pick tomatoes. If you see it there is no danger; you can chop its head off with that hoe."

With much urging Dab lifted the hoe and struck once, twice, thrice and then called out, "I got 'm; 'e daid!"

"Bring it out! Don't leave it in the weeds!" I said.

Dab lifted his hoe tremulously, and there was a small ribbon snake, a foot long and one inch round!

I could not help a burst of merriment over it — and that restored our nerves. Dab continued to declare that the snake had sung, and since, I have felt I was very stupid not to know that the little snake's cry, if snakes ever do cry, was one of terror, and that it was due to the big snake being near, and that if I had only known it was not the monster Dab saw, and if I had not let him waste time on the little snake we might have caught up with the big fellow, who will now remain a permanent terror.

I am going to turn the horses in that field and the cows, and it will be a miracle if none of them meets him, and then my beautiful red setter will always be in danger. However, there was nothing to be done and I went on through the grass to the hay-field, walking very warily ahead with the hoe lifted, while Dab followed in my wake.

We picked nearly a barrelful of cockspur roots from the field. I have had an empty barrel put there to receive them. The peas are bearing well and the grass is very high, and it will make splendid hay, but I will not mow it until I feel sure there is not a single cockspur left.

They are fatal to horses. So strong are their little barbed points that if swallowed they pierce the intestines and kill the animal. There is only one way in which they can be got rid of, and that is by my all-day presence in the field, so for a week I expect to give myself up to it entirely — huge straw hat, blue denim apron, and buckskin gauntlets.

September 21.

This morning I went early to Cherokee and drove through the "street" to get some hands to break in two acres of corn which, being very near the road and convenient to passersby, had better be in the barn. At the well I found a picturesque group of gossiping matrons. After the usual civilities, I told my errand. "Becka, I want you," I said to one, a splendid figure, who stood balancing on her head a large tub of water. She answered: "Miss, I berry sorry; I kyant possible cum, I got de feber right now," and she walked off at a swinging gait. I turned to an equally fine specimen of

One or two hands in the barn-yard.

health and strength and said, "Agnes, you will come?" "Miss, I too sorry, but mi baby got de feber;" the said baby looked as bright and hearty as the mother. All through the street it was the same thing. One elderly woman, quite as a favor, went home and locked her door and came. I had brought my house servants to help and found one or two hands in the barn-yard; but it took much longer than it should have done.

This corn had been stolen in a very clever way. About a month ago I went through the field to mark what I wished kept for seed from the stalks that had more than one fine ear. I found that about every eighth stalk had two ears and some few had three ears; to-day, when gathered, not a single stalk had more than one ear. In spite of this and the damage from the storm, these two acres made seventy-two bushels of shelled corn, which is a comfort.

On the way down I stopped at the post-office. While I waited for my stamps a negro drove up and took from his buggy two large sacks stuffed full of something; each sack held two bushels. To my amazement, when he proceeded to empty the contents on the ground, I found they were rice birds! I tried at once to buy a dozen, but he said they were already sold, and began to count them out to another negro. He had got to 150 dozen when I left and had not got through with one sack. He said he got 35 cents a dozen for them. I have only had rice birds twice this season; yet the fields are swarming with them.

The work of repairing the screw which carries the rice from threshing mill to shipping barn is nearly finished. It has been very expensive, and my crop this year does not warrant the expense. Yet it was dangerous to leave it hanging as it was, and so I was forced either to pull it down, which would have been an expense, or repair it, and I chose the latter course.

PEACEVILLE, September 23.

Went to Casa Bianca to-day, but did not see Nat, though he always assures me that he never leaves the place for an hour. In spite of the rough preparation of the ground the peas I had him plant are splendid.

I went down especially to see the spot I have enclosed in wire, intending to try celery on it. I gave Nat very special directions about preparing the land, but thought it best to see how he would succeed before risking any money in plants. I told him to plough it once north and south very deep — I was willing for him to do only half an acre a day so as to be sure of its being well done — then to harrow it thoroughly and after that to plough it east and west, then to harrow it every day for a week. These seemed to me clear and sensible directions, and I gave him as long as he needed to do the work, not hurrying him.

When I saw the result to-day I was uncertain whether to laugh or to cry; fortunately mirth won the day. I was wearing heavy boots and yet it was difficult walking, so uphill and downdale was it. I am truly thankful I did not go to the expense of buying the plants until I saw the condition of the land. It would be hopeless to expect anything like celery to grow and thrive in such a rough bed; it could never be a success.

A corner of Casa Bianca.

It is a great disappointment. Nat is in some ways so faithful and intelligent that I thought I could make him understand how I wanted the soil. He is a fine rice-field hand. He rented ten acres and always made good crops. This is only one acre of very rich black land with a western slope to a little branch; it has been pastured for years.

In the happy days when I lived at Casa Bianca (about a hundred years ago) it was the vegetable garden, and in it we always grew delicious celery; but then the gardener was an expert, one of the wonderful products of the past, Paul Wynns by name. I should like to tell his story some day. Thanks to his fidelity, cleverness, and diligence the family silver was all saved in the very teeth of the all-absorbing Sherman.

It was some years after the war, and he was very old when he looked after our garden, having a boy under him to do the work. He was a Methodist preacher of some distinction and had great power with his own people, which was very fortunate, for in a time of upset and intoxication, when the poor darkeys were rudderless and one heard the boast often, "De bottom rail dey on top now," Paul's good sense and good heart — I may say his wisdom — were a great blessing, and he left his mark behind him. In the time before 1860 he was in charge of everything in this household, a most accomplished house servant.

My predecessor at Casa Bianca was a woman of immense ability and cleverness. She spent much time abroad and was a great friend of the Grand Duke of Weimar, who on one occasion about 1862 said he had always desired an African in his suite. Mrs. P. said at once: —

"I will send you one as a present."

The Grand Duke demurred, but on her return home, though the war was raging, she fulfilled her promise. She asked Paul if he would like his son Tom to be the lad chosen to go, that he would have the best education and live in the midst of luxury. Paul, after mature deliberation, accepted the honor for his son and in spite of war and turmoil Tom was sent.

The Grand Duke was delighted with him and treated him with the greatest favor. He married the daughter of an "honorable Councillor" and lived happy ever afterward. He lost his life in his efforts to render help when a fire broke out in the palace, dying from the effects of overexertion. His monthly letters were the delight of his father. Since Paul's death I have heard nothing of the family.

When I got back to Cherokee at 4 o'clock I found a funeral going on. David's eldest son was buried. I am so sorry; he was always a good boy and had learned the trade of carpenter and was doing good work. It is hard on his parents.

Elihu's little boy was also buried to-day. I am distressed for poor Elihu. He has lost his wife and three little boys since he left Cherokee. If I only had an empty house in repair I would insist on his coming back. They say it was his poor wife who persuaded him to accept the offer of my neighbor.

As I drove home to-day Ruth shied violently and, looking down, I saw a terrible looking black man in the broiling sun in the ditch asleep or ill, I couldn't tell which, but Dab stuttered out: "Drunk, ma'am; nothing but dat." I drove on a little way and then said: —

"Dab, that poor creature will die in that burning sun. Take my umbrella and go back and set it up over him. Don't speak to him, just put the umbrella so as to keep the sun off."

So Dab flew off, but Ruth would not wait, and I had to drive on. I met a nice looking black woman whose parents had belonged to us, and I said: —

"Chaney, I sent my umbrella to put over a man in the ditch there; do fix it right when you pass."

She dropped a deep curtsy and said: "Dat is my husband Jupiter, Miss Patience, en' he's drunk all de time, en' I t'ank yo' kindly for puttin' de hambrellar ober him. Miss Patience, he ain't gi'e me so much as a apurn fo' five years, but he is my lawful married husband an' I bleeged to ten' 'um."

September 29.

Vareen harvest begun, a perfect day, the sun in great glory, with little white clouds flitting hither and thither, doing continual homage to him, and making the sky a thing of beauty. I did not go down to the plantation early, but followed my plan of getting there just in time to turn back the hands who are leaving the field with too little done. Yet they got ahead of me, for they had all left the field and gone home at 11 : 30 o'clock, having only cut four acres in a field of eleven acres. Of course it was vain to attempt to get them back. I met faithful old Ancrum, whom I had put in charge, and he told me that they had all cut what was counted a task in slavery times, and left the field by 11 o'clock. I was greatly tried, because the risk of leaving the rice in the field all day Sunday is too great, and I wanted to get it into the barn-yard Saturday evening. I explained this to the old man and told him we would have to get a big day's work done to-morrow, as so little had been done to-day, or it would leave a very heavy day's work for Saturday,

"Chaney."

which they all dislike very much. My father always allowed a very light task for Saturday and required that washing, scouring, raking the yards and burning trash should be done in each household as a preparation for Sunday, when everything should be tidy and clean. They keep up the practice very generally now, and it is rare to find on the "street" a house where active preparations are not being made on Saturday evening, and I encourage it in every way in my power.

The new beater for the threshing mill engine has arrived and is being put up. Last year I lost my engineer, he having been absorbed by a neighboring mill-owner, and I felt much at a loss, but I turned at once to an old "befo' de wah" darkey, who had learned his trade under my father. Every one said old Tinny could not possibly run the mill: he was too old and stupid; but I sent for him and he came promptly, and when I asked if he could run the engine and thresh the crop for me he answered, with great spirit, "Suttinly I kin," as though I had insulted him by the question. He has showed himself a competent engineer, careful and vigilant, though he looks as if he had not intelligence or capacity enough to kindle the fire. His first action was to tell me, after examining the machinery, that I must get a new beater, as he did not consider the one in use safe. When I demurred he said, "Miss, lemme mek you sensible. I kin patch um up en run de ingin ef yo' kyan't possible buy a new one; but it's a resk, en my ole marsta 'ood neber expose none o' him peeple to run a ingin wid sech a beater, yo' onderstan', ma'am?" I needed nothing more than that, and wrote at once to beg Capt. L. to come and examine it and, if necessary, to order a new one for me. He took a long time to come, being a very busy man, but when he did come he said Tinny was quite right and a new one was necessary, and now Tinny is engaged in putting in the new heater. It seems almost a miracle to me that he should be able to do it; but it just shows what it is to have been thoroughly trained to a thing in youth. This pygmy of 75, who has not looked at an engine for thirty years, and has just lived under his own vine and fig tree and worked his own little farm, the moment he is called upon, is perfectly at home in the engine room and really more competent than the very intelligent, smart young man I had before, who reads, writes, and speaks correctly and has learned his trade since the war.

I

In the same way old Ancrum, who is 80 years of age, is the one man I can get to do a really pretty piece of ditching. Auerbach says, "By work we learn fidelity," and I believe the immense number of infidelities, financial, moral, and spiritual, which flood the country come in great measure from the sentiment against labor which has crept over the land with the rise of wealth. There is a sentimentality which is opposed to work and laments over the necessity for it, whereas the man or woman who has never really worked is to be pitied, and will never reach the point of excellence and development that could have been attained, had he or she learned to put out the whole strength, either of mind or body, on something.

September 30.

I got down to the plantation in time to turn back some of the young men who had left the field and were on their way to "the street," having cut a half acre but not tied up the rice they cut yesterday. A few laughing words as to the contrast between their strong looks and feeble deeds made them turn back, and fearing to lose sight of them I offered to take them back to the field in my boat. If I had been in the field all morning I could not have kept them, they would have slipped away from me just as they had done from the foreman; but arriving fresh and cheerful on the scene I can force them back by my will. I got into the field just as they had all finished cutting and were about to leave, and as each one turned to leave, I said: "Now tie up what you cut yesterday and tote it to the flat." It was just touch and go as to whether they would flatly refuse or obey. For one moment they stood wavering; then I said, "Don't delay now, for it is better to have the extra work to-day than on Saturday." That settled it and they flew, and now, at 2 o'clock, the whole of yesterday's cutting is in the flat and every one is gay and happy.

Agnes has just passed me going home. As she was getting into her boat I said, "Finished already? I know you are glad I made you do it." She showed every one of her perfect teeth and said, "Miss, I too tenk yo' for mek me do um; to-morrer I kin finish by 10 o'clock." I brought a basket of beautiful Keiffer pears with me and distribute them from time to time, and they are much enjoyed. This country is the home of the pear; both the Keiffer and Le Conte grow and bear luxuriantly, and the pears reach immense size.

I feel so happy at the success of the day's work that I am going to eat my frugal meal, with its accompaniment of artesian water, with great enjoyment. No one who has not spent days out of doors, with all the pretty sights and sounds which nature so lavishly provides, can know the exhilaration I feel. After trying everything for lunch I have settled on a closely covered dish of rice, which is most satisfying and is very little trouble to eat. If only the field did not smell so terribly! My good Chloe has put up a large supply of rice and broiled ham to-day, so I am able, after I finish, to offer a part to any one who looks dejected or tired. "Would you like some of my dinner, Ancrum? Well, bring your bucket cover." They all carry their "bittle," as they call their lunch, in bright looking tin cans with close fitting covers which make nice plates.

When the rice was all nicely stowed in the flat I got into my boat and came home.

October 1.

A sparkling welcome to October — a perfect day with mercury only 65. I am sitting on Vareen bank watching the "toting" — such active, wonderful figures, I wish I had my kodak. The distance across the field is considerable and to see little Stella, just her feet to her knees visible, so huge is the bundle of rice on her head, coming across the field, stepping over the quarter drains from one boggy spot to another, is wonderful.

The hands have worked splendidly to-day and my little refreshments have been much appreciated. Fortunately it was just high water at 3 o'clock when the last sheaf was put in the flat and so it could be poled up the river and put safely under the flat-house. I put Elihu in charge of her as watchman until Monday. I hope that, as the rice in the flat will make a comfortable resting-place, he will remain at his post. It was with a light heart I drove back to the pineland, for the clouds were darkening and it was pleasant to know that the rice is under shelter, and the blessed Day of Rest will be free from anxiety.

October 3.

The first day of threshing is always trying. The feed house is packed up to the very roof with the rice from P. D. Wragg, and I want to get it threshed out to allow Vareen to be brought out of the flat and stowed in the feed room. Of course the belts, etc., all have to be adjusted, and it took so long to get in good running order that when they got through threshing the rice in the mill they all declared it was too late to unload the flat. I insisted, however, on their working until sunset, as they had spent many hours idle while the bands were being adjusted. We got nearly all out of the flat, and it will be easy to finish early in the morning and have the flat empty and ready for Cicero, to whom I have promised it to-morrow, to load up his rice at Casa Bianca.

I rode down on my wheel this morning, a most inspiriting ride in the fresh morning air. On my way to the barn-yard I turned aside to see the field I have recently enclosed, and planted in cow-peas preparatory to alfalfa. There is a splendid growth of peas in full bearing, the pods quite green still. It is a beautiful and cheering sight. I opened the gate and went in, for the finest peas are not visible from the gate. What was my dismay to find ten fat, sleek oxen standing up to their bodies in the peas eating rapidly! They all belonged to the negroes on the place. I never saw a more perfect picture of satisfaction. I walked round the fence till I found the place where they had literally torn three panels to pieces — new American fence wire well stretched on fine cedar posts! I cannot understand it, unless they had help. The top wire had been broken just between two staples and that gave the slackness which enabled them to destroy it. I had just to leave them there, for even if I had not been afraid of them, I could not possibly have driven them out alone.

I had to go on to the barn-yard and not say a word about it until I found some one who could be spared from the threshing — there were just enough hands to run the mill — Jim had gone to Gregory for a load of boards. After a while, in a pause of the threshing, I took Marion, who was stowing back straw in the barn, and sent him with my little Imp to drive the cattle out. I gave him a pencil and piece of paper and told him to write down the number of cattle and the names of their owners, saying, "this is a position of trust, Marion." He answered, "Yes, ma'am," most pleasantly. He came back after a while with the names of the owners and the number of cattle very neatly written, but there were eight instead of ten. I asked Imp afterward how many oxen there were and without hesitation he said, "ten"; so I knew Marion had failed in his trust. Later I had the fence repaired as best I could and told all the men they must tie up their cattle for the night. Elihu, who had three splendid oxen in the field, expressed great regret and said, "I ploughed de lan' for dem pea, en day is tu fyne fer cattle 'stroy." He promised faithfully to shut his up.

October 4.

On my way to Cherokee this morning I stopped at the alfalfa field and there in the midst were fourteen head of cattle; only one man had shut up his. Elihu's three oxen were there and his cow and two pretty heifers besides, also a pair belonging to a man who lives on his own farm, two miles off in the woods, and only works here when it pleases him.

I went on quickly and sent Jim to take Imp and drive them out of the field and into my yard, where the owners can come and pay for them before they take them out. I charged 25 cents each for the first offence, and doubled it for the second. It certainly is a great trial after the heavy expense of such a fence to have professional fence breaking oxen tear it to pieces. I thought nothing could hurt it but tools in human hands.

The fields that have been threshed have turned out pitifully and I am in despair. I hear on every side that the price is very low. Nearly all the planters have already announced that they will not plant any rice next year, which no doubt is wise, but what will become of the country with no money crops? For the first time I put a mortgage on the place this year and borrowed $1000. Marshfield at Casa Bianca (25 acres) has often put that much in the bank and sometimes more; so I felt justified in doing it, but now —!

I am trying to cut and cure some pea-vine and crab-grass hay, but it is very uphill work. Every one is so ignorant of hay making and I cannot tell them with authority because I know nothing myself except what common-sense dictates. The putting up and starting of the mowing machine was very difficult, but now it is working fairly well, and the weather is perfect for the purpose. The stacking I cannot get properly done — they are accustomed to pile straw in heaps and they will only pile the hay instead of making a compact stack.

October 11.

Digging potatoes at Cherokee, eight women with hoes; but they make slow progress. I insisted on having Jim open some with the plough, but Bonaparte said the plough covered up too many, and as he has been superintending this work a long time, and has the banking and storing of the potatoes, I thought it the part of wisdom to let him do it in the way which he assured me would secure the greatest number of potatoes for my use.

October 13.

Still digging potatoes, though only one and three-quarter acres were planted, and they are not turning out as well as usual; they generally yield over one hundred bushels to the acre. The hay making goes on pretty well. Jim is getting to run the mower and rake very skilfully, but the man I have stacking the hay is very obstinate. As long as I stand and look at him the stack is packed and properly formed, but as soon as I leave he just tosses the hay lightly on the pole and a rain would ruin it.

October 23.

The potatoes are all in and the hay nearly so. The other evening I was superintending the stacking of the hay when five children came to ask me to let them go in the potato field and "hunt tetta." I let them go, as I always do, for my heart is tender to children and I like to see their delight over the potatoes they find. I was so much interested in getting a perfect stack that I went up the ladder to the top to see if it was well packed, while the wagon went for another load. It was so lovely up there that I sat a long time. The sun was nearing the end of its journey, and the slanting rays glorified the fields with their borders of bright colored leaves, the ruddy brown of the cypress giving its rich tone to the landscape. I saw from my vantage point nearly the whole upland, and in the foreground the children in the

120   A WOMAN RICE PLANTER

potato patch. They all had hoes and it struck me they were digging very regularly in rows and not here and there, as they generally do, and I watched them more closely. In the little time that they had been there the boys had each about

Five children asked me to let them "hunt tetta."

a bushel in their bags, and I realized that the women had systematically covered up potatoes in the rows as they dug them. I did not stop the children, but let them go on every afternoon, with the result that they each got about ten bushels of potatoes. Another year I will not employ the women on the place to dig them, but will get hands from outside, for whom the temptation will not be so great to hide the potatoes for their own children to find. I like the children to glean, for their parents are so careless and improvident that very few make a crop of potatoes, though they have every opportunity to do so, and children always love potatoes; but when it comes to having the best ones covered up for them I feel it is time to call a halt. One year I superintended the digging very closely myself, and there was no chance for covering up. The crop turned out finely and I was pleased, but after the potatoes were banked in the barn-yard they were stolen, so that I have since left it entirely to Bonaparte.

October 31.

The harvest of my twenty-five acre field at Casa Bianca began to-day — most beautiful weather and the hands worked very well, cutting down seven and a half acres, so that I hope we will get it all in the flats by Saturday.

November 1.

Another brilliant day and the hands getting on merrily with the work. If this were April rice we would tie up to-day what was cut yesterday, but the June rice straw is so green that one day's sun is not enough to dry it and so the tying will not begin until to-morrow.

November 2.

Seven and a half acres cut again to-day and Monday's cutting tied up and put in little cocks in the field. Though we have only had the few hands living on the place, the work is getting on finely. The sky is somewhat overcast, but I trust it does not mean rain.

November 3.

It began to rain late last evening, and poured all night. I could not sleep for thinking of my rice on the stubble. That

"It is tied into sheaves, which the negroes do very skilfully, with a wisp of the rice itself."

which is stacked may not be much hurt, but that lying untied on the stubble will be terribly injured. During all the beautiful weather of the past two weeks I was eager to get the field harvested, but Marcus said it was not quite ripe enough, and when rice is cut underripe the grain is soft and mashes up in the pounding, making a very poor quality of rice; so I was forced to wait.

November 4.

Reports from Casa Bianca are terrible. The gale of east wind we have had forced in the sea water till it swept over the banks, and only the tops of the stacks are to be seen above the water and it is still raining. Marcus had to put a boat in the fields and he paddled down over all the banks to examine the condition of the rice in Marshfield.

November 7.

To-day I moved from the pineland to the plantation (Cherokee). There has been no ice, but we have had three heavy frosts and I think the vegetation sufficiently killed to make it safe.

November 10.

A glorious day after all the rain. I have not written for some days because things are too depressing all around me. When they get very bad I cannot bear to write them down. Saturday I paid out $75, the amount it usually takes to put Marshfield in the barn-yard, and it is still in the field. The turning and drying of the rice have been very expensive. To-day I went down and was much relieved to see it in such good condition. Marcus greeted me with that subtle flattery of which the darkies are masters, a cheerful, respectful, hearty greeting and then, "Miss, de Laud mus' be love yer, ma'am! I neber see sech ting, I was shock wen I see de rice, fu' it ain't damage none tall, yes, ma'am de Laud must sho'ly love yer!" I expressed my gratitude for the great mercy, for indeed it looks wonderfully well. One flat, the *Sarah*, was loaded to-day. She was to have had eight acres put in, but when they got seven on she began to leak and no more could be put on. I have ordered hands down from Cherokee to bring her up the river by hand, for she is

"The field with its picturesque workers."

leaking too much to be left loaded until Saturday, when I have ordered the tug to tow the others up.

November 15.

Down at Casa Bianca again, in the field all day, the hands toting rice to *78*, my largest flat. She is expected to carry nine acres. It is lovely down on the banks, and my English friend, an artist, who is sketching the field with its picturesque workers, is enthusiastic over the wonderful soft colors and the enchanting haze over all. I will have to borrow a flat, for *Sarah* is leaking too much to be brought back from Cherokee and *78* and *White House* cannot carry all the rice.

November 19.

The tug brought the three flats at daylight this morning. I could not get all three unloaded, but the rice from two is safely stowed in the mill and the other will have to take its chances in the flat till Monday. The hands worked well to-day, and were very merry and danced for my artist friend. A man came bringing $2 to buy two wagon loads of rice straw. It is in great demand and it is hard to refuse to sell it when people want it so much. I let this darky have the two loads. I have always given away a great deal but I have to deny myself that pleasure this year, for I have twenty-eight head of cattle, not to speak of the horses, to get through the winter, and the crop is so short.

November 20.

Marshfield turned out $737\frac{1}{2}$ bushels in spite of storm and salt. Now, if I can only get a decent price for it.

November 25.

Drove down to Gregory to sell my rice in the rough, as I have not yet got samples of that I sent to mill in October. Sold it for $42\frac{1}{2}$ cents per bushel, $313.43 for the $737\frac{1}{2}$ bushels! "Alas, poor Yorick."

CHEROKEE, November 27.

Rode on horseback to Peaceville to-day to get the mail, and brought back a very heavy mail and two books which have been generously sent to the Book Club; and not content with that, saw some very nice salt fish at the store and bought two pounds and brought that home too.

I have given Ruth holiday since moving, and am using

Romola. She is a delightful saddle horse so that I have been riding everywhere instead of driving, and I do enjoy it. Romola has a history.

One of my hands some years ago got into trouble and came to me in great distress to borrow quite a large sum of money. I lent it to him and two years passed without his making the least effort to pay it, though he had made good crops and shipped over a hundred bushels of rice of his own to market. So one spring I said to him,

"As you will not pay your debt yourself, you had better make your horse pay it. I will rent her from you and use her until the debt is paid." He seemed very pleased at the idea and brought his mare the next day. I had often felt sorry for her; she struck me as having once been some one's pet and a pleasure horse — a dark chestnut, with a nice air about her. When I asked her name he gave the name of one dear to me which I could not bear to use, so I said: "I will call her Romola, after you." This delighted him, his name being Romulus, pronounced by his friends Ramblus.

I found to my dismay that Romola was too weak to do any work when she first came and I had the pleasure of feeding her for a month before she could be of any use. Romulus had only fed her, and that lightly, when he used her, which might be once a week or once a fortnight; the rest of the time she was turned loose in the woods to hunt her living.

After being well fed and groomed for a while she became quite useful, and at the end of nine months the debt was paid and I returned her to him. He brought her back, however, at once and said: —

"Miss, she look so fine you kin keep um fu' she feed. I ain't got no co'n. I ain't got no pertikler use fur um."

So I kept her through that winter and in the spring he came to say he had received an offer of $45 for her and he was going to sell her. I told him I would give him $50 and so Romola became mine, and she is a delightful creature.

Having known evil days she appreciates her home and is always cheerful. Her gaits are very pleasant, easier than Ruth's, but she is a great jumper, no fence can hold her, she skims over like a bird. When I try to get her near enough to a gate for my short arms to reach the latch there is always a danger of her leaping it.

She comes up to it nicely and stops where a man's long arm could open it with ease, but for me it is hopeless. I ride off and bring her back two or three times with the same result, then she loses patience and prepares to jump.

Green has given me notice that he wishes to leave my service the end of this month, so I must find some one else. He milks the five cows and ploughs a quarter of an acre of oats a day and thinks he is overworked; told Chloe yesterday he was broken down with hard work!

Just at the end of the war, when things were being adjusted after the upheaval of the Emancipation Proclamation, my mother was trying to arrange a contract which would be just to all parties, so that the lands might be worked and the starvation and want which was threatening this region prevented. The intelligent negroes saw the necessity and gave what help they could, acquiescing in the terms of the contract. The inferior element among the negroes was very turbulent and rebellious and it was a very exciting scene.

At my mother's request a United States soldier had been detailed by the commandant in Gregory to be present, witness the contract and keep order. During the turmoil and uproar the soldier said: —

"I should think you'd rather get white help."

From time to time it has recurred to me with renewed humor, and now I think the time has come when I really must try and "get white help."

CHAPTER IV

Thanksgiving, November 28.

I ROSE very early so as to make the long drive to Gregory in time for church. I sent Chloe and Dab out to collect holly and moss, for my thanksgiving service is always to lay some tokens of loving memory in the sacred spot where my loved ones lie.

The morning was beautiful, but very cold; as the sun gained power it got warmer and the air was delightful. I was detained getting off so that I was late for church, but spent a long time in the churchyard placing the quantity of brilliant holly, the berries so red and the leaves so green, in beds of the solemn gray moss to my satisfaction.

When I finished I drove to Woodstock to spend the rest of the day and night. On my way I saw by the roadside two young people having a picnic *à deux* — a pretty woman, very fair in a Marie Louise blue shirtwaist. I thought what a charming way to pass their holiday, taking their lunch in the woods, the brown carpet of pine needles spread at their feet. As I came abreast of them the man crossed the road and said: —

"I wish to speak to you, ma'am. I've been waiting for you. You may remember you passed us driving in a wagon this morning? The man whose wagon we were in and who was driving, said: 'That's the lady for you; she's got plenty of land and money and you'd better see her.'"

I laughed and said, "He was right about the land, but

much astray in the other statement. I have about a thousand acres of land, but not a cent of money."

"Well, ma'am, it's the land I'm after. I want to farm. I've been working with a big company at my trade, steam-fitting and carpenter's work, and they've laid off their hands in this tight spell, and I've took a notion to go back to farming for a while. I was raised on a farm an' was a-ploughin' cotton when I was 12 years old — I don't belong to this State. I come here last year for my wife's health. She loves the country, so I would like to take about thirty acres on shares."

I asked if he could manage that much alone. He pointed to his pretty wife and said: —

"She's just the workin'est woman you ever see an' she'll do her share, I reckon."

I told him to come up to Cherokee as soon as he could and look over the land; that I had a cottage which used to be our schoolhouse when I was a child, which I thought would be very comfortable for him after a little work. I asked him what shares he proposed. He said: —

"In course I don't know the way you works shares in this State, but at home I rents my farm to my brother-in-law an' I furnishes the team and feeds it and the land is under good fence an' we divides the cost of fertilizer an' he does all the work an' we shares the crop in half; he takes one-half and gives me one-half."

I told him that would suit me entirely. I had my land under good wire fencing and would furnish a team and feed it.

I drove on — I have always said I was the special child of Providence and here is an instance — waylaid on the road by the very person I was wanting to find and have been looking for in vain.

I was late for luncheon, but was forgiven in view of such unforeseen interruption.

K

WOODSTOCK, November 29.

This morning it poured torrents, so I did not start until midday, when it was not raining so hard. I drove through the terrific neighborhood road to the ferry only to find the wire broken and the flat drifting down the river.

In the intense cold and wet discomfort, I had food for devout thanksgiving that I had not been a little earlier and so been in the drifting flat. I turned and drove three miles up the river to another ferry, so that I did not get home until very nearly dark.

When within a mile of Cherokee I met my farmer on his way back to town; he had hired a horse and gone up to look over the land, and though it was a most discouraging day and he was wet to the skin and very cold and very sore, for he said he had not ridden for years, he was delighted with the land. He said, however, he feared the repairs on the house would cost more than a renter for only one year would pay, and that was all that he now proposed to rent.

I told him I was willing to put the repairs in and that while they were going on he could occupy two rooms that I had elsewhere, as he expressed great eagerness to come at once if he came at all. So there on the road in the rain, it was agreed that he should come up on the boat next Wednesday.

I am so worn out with the long drive and the intense cold that I can scarcely make myself write, but apparently my "white help" is in sight and I must record it.

December 3.

The boat blew very early yesterday morning. I had sent the two wagons up to meet Mr. and Mrs. Z. and their belongings, and they arrived with very neatly packed clean new furniture, his fine tool chest being the most impressive thing.

Mr. Z. very soon got everything in position and the cooking stove up and going, and this morning he started work upon the cottage.

Fortunately I had some shingles on hand or I could not have undertaken it, but only 1000 will have to be bought. The plastering is down, and that is the most serious consideration now. The sides are good, but the ceiling is much broken.

I drove Romola to the store to get the nails, etc., which were wanted, and then, feeling very much lulled and soothed by the thought of having some one who worked with such vim and needed no looking after, I spent a delightful, restful evening reading the "Memoirs of Madame Vigée Le Brun." Most interesting and inspiring to read of such a woman — such great gifts and above all such wonderful diligence — not an idle moment did she allow herself; her art and the social labors belonging thereto occupied every moment.

CHEROKEE, December 5.

I had to go to Gregory to-day to get the check for my rice. Small though it is, I need it to pay for thrashing, etc. I determined to take my colt Dandy over the ferry for the first time, as that would give a spice of enjoyment to an otherwise trying day, so had the pole put on the buckboard and Ruth and Dandy put in. He drives charmingly in double harness, but the ferry is a very trying thing at first to a horse — just a long, flat boat, only wide enough to admit of driving in with care, without railing front or back, and propelled across the Black River, which is very deep, by two negroes pulling on a wire slack enough to allow the passage of tugboats and small steamers. If, by chance, one of these comes puffing along while one is in the flat, it takes a very sensible horse to stand it.

My horses are all wonderfully intelligent and understand a reassuring explanation accompanied by a pat and loving word, but Dan is so young and frolicsome that he might not stop to listen. He is a picture pony, with the grace and activity of a kitten, and as plucky and stanch as possible, but terribly mischievous; has killed two calves for me. He is not yet broken to saddle, for I was afraid of putting

"The Ferry."

much weight on him while so young. Breaking him to double harness has been a great pleasure to me, for he has never given any real trouble. I put him first in a very light vehicle with Mollie, the doyenne of the stable, who, though old (22) and reliable, is very spirited and pulled up with him beautifully, yet didn't mind his prancing and dancing. I didn't put him in single for fear he would come into general use before he was old enough to stand it.

One day Jim came to me and begged me to allow Jack and himself to put Dan in the little single wagon. I hesitated, as I was too busy to go and see it done, but Jim was so eager for it that finally I consented, told him to take the body off of the little wagon, leaving only the running gear, which would be light, and told him only one of them must be in the wagon at a time. I did not go out for about an hour, when I saw Jim leading Dan to the stable, no wagon to be seen anywhere. I asked where the wagon was. His answer was: —

"Dan went beautiful, ma'am, an' we drove him all over the plantation." "Well," I said, "then, where is the wagon?" Most reluctantly Jim went on: "Then, ma'am, Jack an' me thought as he'd done so well we wud jes' take him down the avenue an' haul in that wood by the gate." "What," I cried, "that heavy oak wood?" Lower and lower went poor Jim's head. "Yes, ma'am." "And what happened then?" I was determined to extract the whole story, so as to know how to act. "Then, ma'am, Dan he pull fine till we cum to rise the hill, an' then he wudn't pull the wagon up." "Did Jack and you take off some of the wood, and one of you push behind?" "No, ma'am, we never thought of that, but we tried to make him pull it, an' when we whipped him he just pranced and threw himself down till we had to take him out for fear he'd hurt himself."

I was very angry. Nothing more injudicious could have been done to the dear little beast who up to this time had thought human beings all powerful and all wise. "Take him back to the wagon, Jim, but give Jack time to run ahead and take off half the load; and put the logs entirely out of sight, Jack, so that Dan may not know that any change has been made in the wagon."

Jack ran ahead and Jim followed with Dan, I walking by him patting and shaming him by turns, and assuring him that he had lost his potato for that day. The wagon was halfway up the steep ascent in the avenue, the only little rise for miles in this flat country. It is hard to believe that those two men had put a genuine load of wood on that wagon, but they had, live oak, which is heavy and strong as iron.

To make things worse, the horses were all loose in the

park, and Dan whinnied after them and they answered. While Jim was putting Dan in I called Mollie and had her halter put on and kept her near Dandy. He stood quietly until Jim took up the reins and clucked to him, then he reared and plunged and bucked, but I made Jack push behind, so that gradually the top of the hill was reached, and then I led Mollie ahead in the direction of the stable yard as though I had forgotten all about Dandy, but told Jim to use the whip freely if necessary, for that wagon had to be brought into the yard by him or he would be ruined. Jack must push behind with all his might so that the pony should not be strained, but come he must.

Jim and Jack both pleaded to leave the wagon till afternoon and then put him in, but I said: "You went against my orders in putting the load on, but having started it you have got to carry it through." Dan proceeded to do all that a kitten would do under similar circumstances — he doubled himself up, he threw himself down, he stood on his hind legs and pawed the air, but finally he leaped forward and took wagon, Jack, Jim, and all up the avenue and into the stable yard at a full run. Mollie and I just cleared the road in time, but nothing was broken, and Dan was in the sweetest humor and no harm was done, for I drove him in double harness the next day and he was quieter than usual; but I have not allowed him put in single harness again, for I want him to forget this episode entirely first.

To return to my trip to Gregory — I started at 12:30, Dan and Ruth in fine spirits and quite playful. When we reached the ferry the man in charge begged me to take the horses out and let him roll the buckboard in and have the horses led in, but I was not willing for that. I have to cross the ferry whenever I drive to the railroad, and my horses must learn to go in quietly, for I often cross without a servant. I had Jim walk ahead and stand in the flat at the point where the horses should stop and then I drove in. The water showed between the flat and the shore, a moving streak of light, which Dan examined carefully, and then snorted. As I touched him lightly with the whip he made a flying leap into the flat and stood perfectly still for a moment, nostrils distended, ears erect like a bronze horse. Before he had time to realize the situation and that we were moving, I slipped out and went to him with an apple and a few sweet potatoes, which he loves. As he smelled them in my hand he relaxed his tense aspect and in a few seconds he was eating as contentedly as though he had been accustomed to a flat daily.

On our return trip he went quietly into the flat and turned his head at once to see if I was coming with a potato, and I do not think he will ever give any trouble at the ferry in future. It was wonderful to see how Ruth did all she could to assist in getting him in quietly. I think she remembered her own first trip, how frightened she was and how I calmed her in the same way with sweet potatoes.

I got through all my business and got back to Cherokee at 5:30, which was, I think, doing well for Dandy's first long drive, thirty miles and the ferry, and he was just as gay in the last mile as he was in the first.

December 19.

Punch came to-day to ask me how much he still owed me. It was hard to tell, for two years ago I sold him a fine plough horse for $50. He had just moved on to my place; wanted to rent land and plant corn and cotton. I heard he was a fine ploughman and his wife a good hoe hand, and I was quite cheerful when he said instead of hiring an animal if I would give him a chance to pay for it out of his crop he would like to buy this horse from me. I had more horses than I needed and readily consented to omit any cash payment and wait until the end of the year.

At first the new broom swept very clean. Punch worked hard and his wife was very stirring and I was delighted; but as the spring ripened into summer and the days grew long and the suns hot, and I moved to the pineland, Punch and Judy began to rest in the shade of the big trees in the pasture, and the weeds grew apace in the crop, so that when the autumn came the results were pitifully small, and I did not exact the payment of the debt, but told Punch, as he was an expert at shingle making, he could cut shingles in my swamp, where there was plenty of cypress, and pay his debt in that way. This proposal seemed to delight him, and he promised to go to work at once. But at the end of two years he had only paid $27 on his horse, and no rent at all for the land he had planted, and he ceased to feed his horse during the winter, so that it died.

His wife was very stirring.

He was in great distress, and in view of his misfortune I forgave him the debt and urged him to work his crop this year. He promised renewed effort and I hoped anew. About midsummer he came to me in terrible trouble. His boy had been arrested and put in jail. He was a boy of about 18 years; his son, but not his wife's; but she in the kindness of her heart when she heard that the child was neglected and starving, took him when 2 years old and cared for him as her own, and had brought him up more carefully than most. The boy had hired a bicycle in Conway, fifteen miles distant, for three days, and had come to visit his father and remained three months. The owners of the wheel had great difficulty in tracing him, but naturally when found they put him in jail.

Punch and Judy, anguish stricken and weeping, came to me for help. I told them the only possible way to help the boy was to let him take the punishment the law decreed. It might save him from being a confirmed thief. All in vain I talked; they pleaded with me, weeping, to lend them the $15 they needed to get him out. They had neither of them slept in their bed since the news first came; they could not go to bed knowing he was in jail. When I asked where they slept they answered on the floor, without mattress or bedding of any sort, and they looked it. Judy said: "Miss, yo' tink I kin git een my comfutuble high bed en kno' dat chile, my own boy I raise, is punish een jail. No, ma'am, I tell Punch neber will I git een dat bed agin till my boy is save."

Unfortunately I had the money in the house, and I gave it. They had sold one of their cows and got the other $15, and Punch went and paid the $30 and the suit was dropped. No sooner was the boy free than he was arrested again for robbing the post-office, and then their disappointment and distress was so keen that they became silent. Judy only said to me: "Miss, I wash me han' of de boy, now; me heart is broke."

It was pathetic in the extreme. I tried to encourage Punch to do some work and pay me in that way, as he had promised, but in vain. I needed shingles more than ever, but with all my efforts he still owes $10 on this last debt. Now he came to tell me that he was going away. He put it with great delicacy and began by saying, "Miss, I dun'no how 'tis, I kyant please yo'; I try en I try, en somehow I kayn't cum it. Yo' kno', my missis, a man kayn't do mo' dan 'e kin. Man p'int, but God disapp'int."

I could not help laughing at this new version of "L'homme

propose, et Dieu dispose." "Oh, Punch," I said, "I think you have got the wrong end of that. I think in this case it is God who points and man who disappoints; but certainly you can go, only you must do something to pay me that $10 before you leave, for I am in need of the money, and I have waited on you as long as I possibly can. I am perfectly willing to take the shingles, and it would not take you long to pay up the debt." It was in vain, unless I had the Sheriff take his cow, which I could not bear to do. He said he would pay it by degrees next year, and I was so glad to have him go, that I gave up the effort to get anything from him.

The two acres of cotton he rented were very near the field I planted. He and Judy did not work theirs, so there was a fine field of grass and weeds, with a few stalks of very tall cotton. Notwithstanding the rarity of the stalks in their bed, day after day I met Judy coming out of her patch with an immense bundle of cotton on her head.

Day after day I met Judy coming out of her patch.

Jim would grow furious when we met her, and now and then break out: "I work yo' cotton an' keep it so clean, f'r Punch pick." There is no doubt that he was right, but no one could ever catch Judy anywhere but in her own patch, where the same few bolls of cotton showed out every day. Jim begged me to send them off before another crop season, so I am glad to have them go, only I do wish I could have got my money.

December 17.

The Zs getting on finely. She is a wonderfully capable woman, and I think a very nice one. She seems so pleased to be in the country again and is eager to take the milking — wants me to send off Gibby and let her milk.

I told her that she could take the Guernsey cow up, that as soon as the calf got big Gibby said she was dry and he could not get any milk, but that I knew it was only because he was a poor milker, and I would be delighted if she would feed her well and milk her; I knew she could bring back the milk. She did not seem very pleased, but consented.

She is evidently not a strong woman, and if a bad spell of weather should come she could not go out to milk, and I would just be left milkless. Better go slowly, I think, and not upset things.

I told Chloe to give them a pint of milk every morning and every evening. The cows are not giving much, but then I am not feeding them as I usually do.

The stringency in the money market affects everything. There is no sale for anything — cotton, cattle, horses — I have tried to sell anything and everything, but in vain.

December 20.

To-day I signed a contract with Mr. Z. which I got a lawyer to draw up. He has been very anxious about the signing before this, but I thought there was no great hurry.

He and his wife have been very diligent, working early and late, setting out a new strawberry bed and getting land ready for other things. She has planted celery very successfully and says this land is just suited for it, and wants to try a quarter of an acre in it. They are charmed with the ever flowing artesian well and are arranging little ditches to irrigate in dry weather. Altogether I feel so peaceful and content that it is hard to write regularly.

Christmas Night.

Had a peaceful, happy day, many loving tokens of the blessed season of good-will. It is always a pleasure to make the darkies happy with small presents and I included the Zs in my offerings of good-will.

Besides many little things to eat I presented them with a pair of Plymouth rocks, a beautiful pullet and cock, as they are anxious to start a poultry yard. This afternoon she came in with an offering for me, a necklace of fish scale flowers made by herself, which she had told me the other day she sold for 50 cents.

I was quite touched by it and by her happiness over the fowls. Altogether I feel very thankful that I have found such satisfactory people. He talked to me a great deal to-day and said he would give $1000 if he could get rid of his evil temper. I told him a thousand prayers would perhaps accomplish his desire better than the same number of dollars. He went on: —

"I've been a powerful wicked man. I've shot two men an' been shot twice myself and I've stabbed one man nine times and been all cut to pieces myself, but for two years now, since I met this wife, I've quit drinkin' an' I'm tryin' to live a good life."

I told him I felt quite sure if he earnestly tried he would succeed and that I would do all I could to help him. I felt a little disturbed for a moment, but a full confession of one's sin is often the beginning of a new life, and the idea of helping a man to a higher, better life adds a new interest to the experiment.

January 1.

Sat up last night to see the old year out, the year which has brought us sorrow and distress, yet there is great sadness in seeing it go. In the last moments of the dying year I sank on my knees and prayed that this whole land might be blessed and guided through the coming year.

The day is brilliantly beautiful and we went to our simple little service in Peaceville. Dear, frail Mrs. F. had made a great effort to get to church "to return thanks for her many blessings." Eighty-five years have passed over her, the first half surrounded by all the comforts and conveniences that money can give. She now has the bare necessities of life, no cook and none of the conveniences of modern houses that make cooking easy. She is always cheerful, always dainty and beautiful to the eye, and one never hears of what she lacks or needs, nor of the possessions of the past.

To-night Chloe came to tell me Elihu is very sick with pain in his side. I sent her out at once with some tea and milk, a mustard plaster I made, and told her to see it put on. She is always so good and willing. Though it was 9 o'clock and quite a walk to Elihu's house, she went cheerfully. They never have anything prepared for sickness. There is a great deal of pneumonia about and I want to take Elihu's case in time. With all his faults he is one of the best men on the place.

January 4.

I am puzzled beyond measure to know what to do for another year. It is impossible to go on planting rice if it is to sell at 40 cents per bushel. It is an expensive crop, and if one borrows money, as I did last year, at a high rate of interest, and puts a mortgage on the plantation, it very soon means ruin. I have no idea how I am to pay off that mortgage of $1000 this year, but hope the bank will be willing to renew.

Instead of being anxious to have the usual first of January powwow over, as I generally am, I shall do all I can to put it off, for how can one do one's share in a powwow when one does not know what to say? I have absolutely nothing to propose. As far as my seed rice will go I will rent rice land to the negroes, and if I had money of my own I would go on and plant, for it seems to me the complete giving up of the staple industry in a country is really a revolution. Our labor understands no other cultivation; the whole population lives on rice, white and black, especially black. It is a wonderfully nutritious and sustaining food, and if suddenly its cultivation ceases there will be much suffering. Our cattle live on the straw, it being the strongest and most palatable of the straws. My horses will not touch fresh oat straw while there is a wisp of old rice straw to be had; the cows and pigs are fed on the flour, a gray substance that comes from the grain as the chaff is removed in the pounding mill. Mr. Studebake, a great Hereford cattle man, told me that rice flour and pea-vine hay make a perfect ration for cows, one supplying exactly what the other lacks. If rice is given up the cattle and pigs will have to go too.

January 10.

To-day I went down to Casa Bianca to receive Marcus's resignation of his place as foreman. He is going to move "to town," to enjoy the money he has made in my service and planting rice. He has bought land there and built four houses, which he rents out. He is a preacher, or, as he says, "an ordain minister." I have wondered he stayed these last few years, but he has made so good an income that his wife was willing to forego the joys of the town; he owns a horse and buggy, three very fine cows and calves, and three splendid oxen.

I feel very sad at parting with him; he has been here so long, and as foreman he has been most satisfactory in every way. When he turned over the keys of the barn to me I almost broke down, for I hate change anyway, and I really do not know to whom I can give the keys.

King came to beg me to give him a house. He is absolutely worthless and unreliable, but he spoke of his large family and how necessary it was for him to get where he could pursue his business of shadding, and Casa Bianca was the very best pitch of tide for the shad fishing. He gave me an idea, and I told him he could have the house if he would give me two shad a week during the shad season, two and a half months. This he most willingly agreed to do. I never have been able to get any tribute at all from the shad nets, which are set in front of my doors all winter. Five or six men shad there regularly, but they elude all demands, and I rarely eat a shad, as they are too great a luxury for me to buy unless I have company; they are like the wild ducks which swarm in the rice fields at night in the winter, "so near and yet so far."

After much thought and uncertainty I decided to give the keys to Nat; he is willing and knows all the sheep and cattle well, and on the whole is the best one on the place. It is a mere form, for there is nothing left in the barn, but Nat is very proud and happy and the other men very sulky.

January 12.

Cæsar came up from Casa Bianca with Jonas and King to say they could not stand Nat as head man and to indicate that he, Cæsar, was the man for the place. I said to them: "Do you know why I chose Nat? I looked over my book and found he was the only man who for years has paid his debts to me. Every one else on the place has borrowed money when in distress, or got a cow from me on time and left the debt hanging, in spite of my reminding them from time to time that I needed the money; but every time Nat has borrowed money from me or bought an ox he has paid up promptly as soon as his crop came in. Now, this shows fidelity and honesty, and, therefore, I have given the keys to Nat, and if you do not like it you can all leave."

They were dumb at this. Then I asked each one how much he owed me, bringing out my book to verify. Not one owes less than $8, which they have owed over a year. "Now,"

I said, "don't you think I had good reason for choosing Nat to carry the keys?" They looked sheepish and departed.

CHEROKEE, January 13.

Last night at 2 o'clock Chloe woke me to say Mrs. Z. was very ill and Mr. Z. wanted a horse to go for the doctor. She had sent Dab to wake Gibby to go for old Florinda, the plantation nurse, spoken of as the "Mid" or the "Granny," who lives some distance off across a creek. I told her Mr. Z. could take Nana to go for the doctor.

I dressed rapidly and came down. Mrs. Z.'s face was crimson and she seemed unconscious. He was bending over her crying like a child and wailing out all the time, "O God, help her! I know I'm wicked, but spare her!" It was distressing.

Chloe was bathing her feet in hot water and doing all she could. I rubbed her for two hours and applied mustard until the nurse came, and about daylight she seemed relieved. I had not seen how the nurse could be got, but Dab's account was exciting.

He with difficulty woke Gibby, who when he heard there was sickness at the "big house" got up quickly and they went together to the edge of the creek, where they shouted and knocked on a big cypress tree with sticks until the old woman came out of the house down to the edge of the creek, on the other side. When she understood it was sickness at the "big house" she jumped into her paddling boat which was tied there and without going back into the house paddled herself across, and when she landed, Dab said "she tie up her coat to her knee an' start to walk so fast that Gibby en me had to run to keep up."

"Old Florinda, the plantation nurse."

This is an old time plantation sick nurse, who, though now very old, flies to relieve the sick with enthusiasm. She brought herbs with her and soon relieved the patient. This morning I lent the horse and buggy to Mr. Z. to go down to Gregory and consult the doctor.

January 14.

Last night Mr. Z. came to ask me to lend him a lantern every evening. I said I would with pleasure. He said he wanted to pull corn stalks at night, that Maud, his wife, could do it two hours every night and not waste daylight on it.

I said I thought if he worked all day it would be as much as he could do, but he could always get the lantern. He went on in a conversational way to say : —

"I've got a fine burn on them piles o' trash."

"I hope it is well out, Mr. Z. There is such a gale it is no time for burning trash. I hope you saw the fire entirely out."

"No, Ma'am," he said, "I've got it started good, an' it's burnin' fine."

I said not another word, but flew through the house to the pantry, seized the lantern and called to Dab to follow me. We ran at full speed to the barn-yard, where not 200 feet from the threshing mill (which cost $5000) and four large barns three bonfires were raging, the flames and sparks whirling and licking out in every direction up to high heaven, it seemed to me.

There was nothing to be done but watch until the piles burned down. Then I had Dab cover the lightwood posts and beams which Mr. Z. had put on to insure a good burn, with earth.

If I could have got at other hands I would have called them, but it is half a mile to the "street," and there was nothing to do but help Dab myself as much as I could. I had sent him for hoe and spade and shovel, and he worked splendidly.

Mr. Z. had followed me down, also his wife, though I

begged her not to come out, having been so ill yesterday. He would not help in any way to put out the fire and kept saying the wind was blowing in the other direction.

"Yes," I said, "but the wind does not take long to shift, and if it did change there would not be a building left on the place. Dwelling-house and all would go."

I noticed that he got very white as he stood and watched me, but I was too actively employed to watch him, but I thought the tears were running down his cheeks as he stood in the fierce red light. Mrs. Z. hovered around a while talking to him in a low tone and then she left.

When Dab and I got through I had the shovel in my hand and wanted to take the lantern. I handed the shovel to Mr. Z., saying, "Will you take the shovel, Mr. Z.?"

Fortunately, I had the full light of the lantern on his face, and I was shocked; he did not move. I fixed my eyes full upon him and repeated, "You did not hear me, Mr. Z.; will you take the shovel?"

Slowly he put out his hand and took it. I still fixed him with my eye, until he turned and walked toward the house, and I followed him. Dab had gone on before. It was 11 o'clock when I got back.

January 16.

Chloe is in a terrible state of mind, Mr. Z. has frightened her so. Last night he said to her: —

"That missis of yours had a very narrow squeak for her life last night. Twice I had my hand raised to kill her and Miss Z. pulled me back, en at last when she handed me that shovel an' told me to take it I cum as near killin' her right there en buryin' her up with dirt with that same shovel, jest as she had buried up my fires, as I ever cum to anything in my life — en more than that, if she goes to givin' me orders I'll do it yet, en le' me tell you you'd better not tell her this or I'll tackle you. I don't 'low people to fool with me."

Chloe is enough of an actress to convince him that her silence was assured. I thanked her for her confidence and told her she need not be anxious. The fact of the light enabling me to look him in the eye had saved me, and the danger was past.

January 17.

Was busy by the smoke-house this morning when Mr. Z. passed by. He has not spoken to me since the night he set the fires in the gale of wind and I had them put out. He has written me several notes demanding things, to which I have sent verbal answers, and I felt it was time to put a stop to that sort of thing, so as he passed I said in a clear, loud voice: —

"Good morning, Mr. Z."

I was bending over a table at the time, brushing off the hams preparatory to smoking them. He took no notice but passed on as though deaf. I straightened up and said again in a clear voice: —

"Mr. Z., you did not perhaps hear me; I said 'Good morning.'"

He stopped and slowly raised his hat, said good morning and passed on, and I knew I had scored another victory.

About half an hour afterward he came back and said he would like to see me in the field where he was ploughing. I told him I would be at leisure in a minute and would join him in the field.

I went in to get my coat and told Chloe where I was going. She implored me not to go, but I soothed her fears, trying to laugh her out of them. When I got out into the field Mr. Z. asked me some trivial questions about where to plant things, and then he said: —

"You went too far with me the other night, Mrs. Pennington."

"Indeed?" I said.

"Yes," he said. "You told me I had no sense."

"I certainly didn't tell a story, Mr. Z., if I said so. I thought as I stood there and saw that fire swirling around in that gale that I had never seen any one over three years old do a more foolish thing."

We faced each other squarely for a moment. "I saw murder in your eye, but I'm not afraid of wild beasts."

Gradually his face relaxed and I saw the demon had fled for the time, but it was exciting.

After this he talked naturally and pleasantly about what he was going to plant. As I left I said: —

"Remember, you can plant the crops where and how you please, I don't want to be consulted about that, you understand it; but never set a fire burning without asking me."

January 23.

Yesterday being Sunday, I invited Mr. and Mrs. Z. to come in and have service with me, which they did. They went home and made a careful toilet and returned with Sunday clothes, and hats, and kid gloves closely buttoned. I found it a little embarrassing to read the church service, but went through manfully, and a short, simple, clever sermon.

Life has become very interesting with this new problem. I told Mr. Z. the other night that I thought he had better go to my neighbor's who has a nice house in the pineland, and that I thought it would be healthier for his wife, and that of course we could break the contract by mutual consent, but he answered promptly that he did not wish to go anywhere else, that the thirty acres he had taken was the finest land he ever saw anywhere and he was going to make a pile of money for me and a pile for himself; he had been all over my neighbor's land and it did not please him as well.

I wrote to my two lawyer nephews a full account of what had happened, and they both wrote, "For heaven's sake, break the contract!" But I must bide my time to do that. The arrangement was that no money was to be paid at present; all that I owe him for carpenter's work on his house was to be taken from my share of the crop. If I were to break the contract I would have to pay him all that at once, and I have not the money. My cotton has not sold and there is nothing else to look to.

Ran out to meet mail man this morning to get a letter off and found that his horse was quite sick — could scarcely walk. Sent Dab in for the aconite and spoon and gave the horse a full dose, and in a few moments he was able to get on again.

Have had twenty cords of live oak cut and hauled to the river, but cannot sell it in Gregory, as I hoped.

January 28.

Yesterday had Green take Dandy, my beautiful pony, to Mr. F. in Gregory to be sold. If I can sell him now, I can pay my taxes. He is so beautifully formed and so easily kept and so gay and so fond of me that it is a great trial to send him off; he would make a splendid polo pony, but if I can make him pay the tax I must do it, for I still have three grown horses and two colts.

February 2.

Up till 1 o'clock last night with Mrs. Z. She was unconscious for two hours and pulseless for fifteen minutes.

It is dreadful, I said to myself last night as I was trying to pour brandy down her throat and restore her to life. "You poor young thing, if ever you get up again I will try to get you back to your own people." She has four married sisters in her home, wherever that may be; for some reason they do not give clear information as to where they came from.

February 13.

Mrs. Z. told me that she wanted to go home and Mr. Z. is

willing for her to go, but will not go himself, and she is not willing to leave him. She knew he would go right back to drinking and killing people, both of which amiable weaknesses he had given up since they met.

I told her I was not willing to have him stay without her, but not to tell him that, as it would enrage him; just to stick to it that she would not leave him. She gets paler and thinner every day, and I know he cannot hold out. I said yesterday if I only had the money to pay him up in full I would propose to do so and break the contract, and Chloe said at once:—

"Miss Patience, le' me len' yer de money. Ef yu jes send me down to town I kin git um from de bank fer you. Do please, ma'am, le' dem go."

So I spoke to Mr. Z., saying, though it was most inconvenient, if he wished to go with his wife, which was most necessary in her state of health, I would consent to break the contract and pay him the $60 I owed him for work. Most reluctantly he consented.

I sent Chloe to Gregory in the pony carriage, and she brought back the money. I wrote a note for it at 6 per cent, and made her pin it in her bank-book in case of my death.

"Miss Patience, le' me len' yer de money."

February 16.

Paid Mr. Z. up in full for services and gave him a note for his little furniture, and bade them good-by, sending them to Gregory in the wagon with Nana. I felt quite sorry to part with Mrs. Z. She is a nice woman, and, poor thing, married to a madman, to whom she is devoted. Thank heaven he is going and that we part friends. My experiment of white help is at an end!

He took me over all the work — beautiful strawberry bed, with potatoes planted, 900 onions set out, celery bed started, all beautifully prepared. It is sad to think it will soon all be grown up in weeds. I must take up my burden again.

February 24.

In field all day; having oats ploughed in; bitterly cold north wind blowing.

February 25.

In oat field again all day. Gibby ploughing with oxen and Green with Nana.

February 27.

A charming meeting here of the woman's auxiliary. I went out to the oats field intending to get back before 12, the hour of meeting, but Gibby went to burn up some patches of cockspurs and let the fire get away into the pasture, which was terrible. I had to stay and fight it.

I made Green take his plough and make a deep furrow ahead of the fire round in a large curve and had the women beat it out on the sides.

While I was busy with hands and face blackened Dab came running to tell me the "company had come" so that I had to rush home and make a very hurried toilet to open the meeting. We are to sew for an Easter box to be sent to the mountains of North Carolina.

February 26.

At church to-day Miss E. came up and said: "Miss Patience, going to take any one home with you to-day?"

I said "no."

"Well, then, I am going to ask you to take me to spend the night. I haven't seen you for so long."

"With the very greatest pleasure," I answered truthfully.

Miss E. is one of the best women that ever lived and the very best housekeeper to boot. She knows exactly how much to provide for a family of four without waste and yet abundantly, and she can arrange for a table of seventy-five with the same precision; abundance of excellent food and no waste. With such qualities it seems strange that she should have now only the position of what Chloe calls "sextant" to our little village church, and her modest remuneration of two dollars a month is all that she has in the world.

She was a woman of wealth, but, like so many others, her means all disappeared with the end of the war, and she has supported herself by sewing and taking places as housekeeper for a number of years. Now she begins to show the ravages of time and does not feel she can do all that a housekeeper should, and for the last six years has lived in Peaceville, where she had nieces who are devotedly kind to her, but she will not live with them. She lives alone in a house which belonged to her mother and where her summers were spent in her youth. It has passed into other hands, but she is allowed to stay there in the winter as the house is only rented in summer.

It is very near the church, and she is very happy and a marvel of cheerfulness and faith — no repining, no complaining. She sometimes takes in a little sewing still, but for absurdly small prices.

Miss E. is a walking chronicle of the ancestry of every one in the county, I might almost say in the low country, as the coast is called in this State, and can tell you who is who emphatically. I enjoyed having my memory refreshed on many genealogical facts, as I am very weak in that quarter. I am really devoted to this dear old lady and feel it a privilege to have her with me.

February 27.

Drove Miss E. home and then began preparing for the paying guests, who arrived at 2.

March 2.

Finished planting oats at last. I have spent every day in the field for nearly two weeks — the last few days a joy — just drinking in the delicious soft air and watching the buds which promise so much.

There is a mystery of hope over everything, the rest of the ideal, and as I sat on a cedar trunk to-day and looked out into the drowsy blue of the atmosphere I felt a sense of gratitude to the Great Maker and Giver of all this beauty — thankful for my blessings; the great blessing of space and freedom and closeness to nature — yes, and thankful for my limitations, my sorrows, my privations. Thankful that He has thought me worthy to suffer and has taught me to be strong. He is beauty and power and love illimitable and infinite.

March 13.

Jim summoned to Gregory by the extreme illness of his wife, and I have to turn over the stable and cows to poor Elihu, who can't help taking the feed and the milk and is the poorest driver in the world; always touches up the horse that is pulling all the load; yet I am thankful to have him to fall back on. The storm last fall threw down all the pine trees on my 350 acres of woodland and the e are several thousand cords of pine wood lying on the ground which I am trying to get cut and shipped. It has been the habit of many to sell the wood to negroes at the stump, as they call it, for 25 cents a cord. This I am not willing to do, and consequently find it very difficult to get the wood cut. I pay 40 cents a cord for cutting, 30 cents a cord for hauling, and about 30 cents for flatting, and the wood brings $1.50 a cord if it is pine, and $2 if it is lightwood.

The hands are needing work. I have ten men on Cherokee, and if they would work I would have money for all my needs and their families would live in abundant comfort. There is no felling of trees necessary. They are all lying prostrate; it is only to cut them up, and the hauling is only one-quarter to one-half mile to the landing; yet day after day the hands are loafing about the roads, with guns on their shoulders and hide when they see me coming. If I come up on one unexpectedly he is very polite and has some tale of fever all night or a sprained finger or a headache to explain his not working at the wood.

March 16.

Rode out into the woods on horseback with surveyor to get the lines of my land marked distinctly, as all the large timber is being stolen from it by negroes who own lands adjoining. It is terrible to see the trees all lying on the ground lapped and interlaced so that it is hard to get through on horseback.

March 18.

Went out to see the wood which has been measured and is ready to send off.

March 21.

Gog and Gabe have the *79* flat loaded and have sent Elihu with them in charge of flat; they must leave on this afternoon's ebb-tide. I first told Cubby to go with the flat, and he made objections and I got very angry and told him instead to take *Sarah* up the creek to the landing to be loaded to-morrow.

March 22.

This morning a huge lighter arrived, sent by Mr. L. for me to load with wood, but it could not get under the bridge until low water. Had Scipio paddle me up the creek to the landing to see the flat being loaded. Cubby and Sam were loading and they will get off on this evening's tide.

The creek is very wild looking; great trees on each side cast a dense shadow everywhere. Hearing a curious noise of floundering I saw a large alligator crawling through the mud on the edge. He had gone quite a distance from the water in his effort to get the sun, and I had a fine view of him before he plunged in again. They make for the water as soon as they hear a boat approaching. I saw him again as I came back, only for one second, but I saw a number of terrapin sunning themselves on logs. They stretch their long necks and peer with their beady black eyes until the boat gets quite close to them and then drop into the water like a stone with a great splash.

About a month ago I got a note from Mr. L. asking me to allow four negro men to cut 100 cords of wood on my land and he would be responsible for the money, $25. I sent word that I would undertake to have the wood cut for him myself with pleasure, but would not sell it for 25 cents per cord at the stump. I heard afterward that a neighbor had sold them the right to cut on their land, and when I went to the landing to-day I saw about fifty cords of the wood they had cut piled there, and it was the most splendid fat lightwood I ever saw, from trees that had been growing on that land sixty or seventy years. And the owner gets 25 cents a cord, while the wood brings $2 anywhere.

March 23.

Late this afternoon I went up the creek to see the flat that Cubby is loading with wood. The creek seemed darker and more mysterious than ever, as the clouds were lowering and there were mutterings of thunder. The air was perfectly delightful, fresh from the sea.

I enjoyed the expedition immensely until the storm burst, and then Gabriel was unable to manage the boat at all, the wind was so high. I had to get him to retreat to a cove and put me out, and I walked home in a pouring rain, thunder, lightning fierce, and wind so high that it was impossible to hold an umbrella. I am very thankful the loaded flat is up the creek and not out on the river. To-day my new venture arrived — an incubator. I do not see why we could not operate poultry farms with success here, and will give it a trial at any rate.

April 3.

Letter from Mr. L. says the wood sent in three flats only measures up thirty-three cords, when I paid the hands for cutting and hauling forty cords. Fortunately I reserved some money from each one until the wood should be delivered; but another time I will not take any one's measurement but Mr. L.'s, for after it is measured each man carries home five or six logs every evening in his ox cart, and naturally the wood falls short when delivered. I had to do an immense deal of rule of three calculating to find out just how to divide the shortage among them, but succeeded to every one's satisfaction. Live and learn — I will not get caught so again. I spent the morning working in the negro burying ground. Storms have thrown down trees in every direction, and though all the descendants of the 600 who belonged to my father wish to be buried here, not one is willing to do a stroke of work beyond digging the grave he is interested in.

I have told the heads of families that if they will each give 25 cents, which will make enough to pay for a good wire, I will furnish posts and have the fence put up. They seem much pleased at the idea, but I fear it will end there.

I am glad the two marble monuments put up by my father in memory of faithful servants before I was born have thus far escaped injury and still tell their message of love and fidelity in master and servant. The wording is odd, but I think it is a beautiful voice from the past, that past which has been painted in such black colors. Here is the first inscription : —

In Memory of
Joe of Warhees,
Who with fidelity served
My Grandfather
Wm Allston Sen'r
My Father
Benj Allston and Me
Grateful
Whose Confidence and
Respect He had
1840

This was certainly not the gratitude which La Rochefoucauld dubbed "a keen sense of favors to come." The other reads: —

In Memory
of
My Servant Thomas,
Carpenter.
Honest and True
He died as for 40 years
He had lived
My Faithful Friend
1850

It is remarkable that my father did not put his name, R. F. W. Allston, to show who had so honored and remembered his faithful slaves; in another generation no one will know. He was Governor of South Carolina in 1857–1858.

Good little Estelle died yesterday and is to be buried this afternoon, and it was looking to her funeral that I walked through the beautiful spot to-day, and finding so many fallen trees I called Frank to come with his axe and clear it out a little. I can ill afford to pay for the day's work, but cannot bear to have it look so wild and unkept.

April 4.

A perfect day. Last night was so cold that the watermelons, which were up and growing nicely in the little boxes ready to be set out, were nipped.

Chloe returned last evening from Estelle's funeral in a state of exaltation. The preacher had described her death, and it was glorious. He repeated the words she had said: "Yes, I'm goin', don't fret. I'm all paid up fur ebryting. I got um here, right by me, a bag o' pure gold on one side o' me — en Jesus Christ on de oder — en now I'm gwine to de weddin' supper."

Then she asked him to read a certain chapter and at the end of each verse she said: "Dat's it, tenk yu, sah," and when the reading was ended she went to sleep.

Estelle had been our maid for five years and only left us to be married — a good match according to their ideas. She had a new baby every year and worked very hard. She grew blacker and thinner, until early this spring she took to bed. Though scarcely thirty I think, she leaves five living children and three lie in the graveyard beside her.

I never could get her to do anything in the house after her marriage, though it would have been much easier for her to take the lighter housework and with the money hire some one to do the heavier field work. But that is not the proper thing among the darkies of to-day.

A woman may work herself to death in her husband's field, wash, cook, scour, mend, patch, keep house, and receive gratefully any small sums her husband may give her, always answering "Sir" when he speaks to her, above all increase the population yearly — all this is her duty, but it is improper for her to take any service like housework. And so all Estelle's little accomplishments and skill were wasted, except the sewing which I had taught her and that showed in the neat, trim looking clothes of her little army of children. I think she has heard the "Well done, good and faithful servant, . . . faithful over a few things."

To-day two friends of mine were to drive fourteen miles to spend the morning with me. As Dab is strangely agitated and upset by any addition to my solitary meals, I helped him prepare the lunch table before they arrived.

It looked very pretty and dainty, but I saw marks of fingers on my precious hundred and fifty-year-old urn-shaped silver sugar dish, so I told Dab to dip it in hot water and rub it dry with a cotton flannel cloth to remove the marks of his fingers. He was gone in the pantry longer than seemed to me necessary, so I followed him there. To my dismay the sugar dish which he held in his hand looked as though he had greased it thoroughly.

"Oh, Dab!" I cried. "What have you done?"

He looked at me, his face beaming with pride in his work, and answered: —

"I jus' shinin' um up wid de knife-brick!"

Words failed me as I took the precious thing in my hands, but when I had recovered a little I said: "Dab, twenty dollars could not undo the work of those five minutes — no, not fifty dollars!"

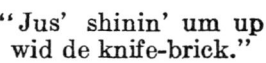

"Jus' shinin' um up wid de knife-brick."

I dipped it into the pan of scalding water and wiped it dry, but alas! no change. Actually the beautiful engraving of little garlands of roses looped around the top was almost effaced, so vigorously had Dab employed those few moments.

Alas! alas! zeal without knowledge is a terrible thing. Poor Dab cannot possibly do just what he is told; he has to plan some original course for himself.

I went to meet my friend unduly agitated and upset by the circumstance, but was careful not to speak of it. I can bear things so much better if I do not mention them to any one until the pang is all gone. That is why this little diary is so much to me. I can explode into it, and then shut my teeth and bear things.

Unpacked the incubator to-day with Bonaparte's help and began to study its mysteries. We had a time getting things right, for he has never seen or dreamed of an incubator, and disapproves entirely of the effort to take away the occupation of the hen and defeat nature, so that his manner was disapproving, not to say forbidding. My good Chloe, too, feels that for some unknown reason the Great Father has given me over to the temptation of the Evil One, and walks past the "'cubator," as she calls it, with head high and firm tread; her manner is what the "nigs" call "stiff" — that means distinctly rebellious and unconvinced. I had only seen an incubator myself for five minutes under the rapid flow of words from the young man exhibiting it, words of fervid praise and faith which left me somewhat vague and confused as to details, for it was just in a shop and not working.

I calculated when I bought it that I would have time to try my 'prentice hand with fowl eggs, which take only three weeks to hatch, and then fill it with turkey eggs, which take four weeks, and get them out before I have to leave home on May 8; but unfortunately the steamboat was detained by a storm and so the incubator was delayed a whole week, which threw out all my plans, and I will have to give up the turkey eggs. The little book, which is wonderfully explicit and satisfactory, says one should study out the management of the heat thoroughly before putting in the eggs, and that will make some delay.

April 6.

I have sat on a low stool in front of the incubator day and night since it was unpacked and installed in the drawing-room. I lighted the lamp at once, and then watched the thermometer, which necessitates a bright light and a very low seat. I thought it was going to be very simple, and on the second day I thought I had it steady at $102\frac{1}{2}$ degrees, and went off into the field to see after some ploughing. When I came back I rushed in to see if it was holding its own and found the mercury at 110 degrees — one little step more and it would have broken the thermometer. After that I just stayed there. The thermostat is a wonderfully delicate piece of mechanism and I have no one to consult.

April 7.

At last I have got the thermometer to remain steadily at $102\frac{1}{2}$ for ten hours, so to-night at 6 o'clock I put in the 120 eggs.

April 10.

Tested eggs to-day. Only six infertile. The thermostat is working beautifully and the mercury does not vary a half degree during the twenty-four hours. I am very careful to follow absolutely every direction and let no one touch it but myself, for I wish to give it a fair trial. All my friends in the county are confiding to each other their anxiety over my venture. "Such a pity dear Patience should have wasted her money on such a folly. A huge sum, $25, for those two machines. It is distressing." Many years ago, when incubators were first invented, a progressive neighbor invested in one, and the lamps exploded and a serious fire resulted, so that it is only natural that incubators are much looked down on in this community. No doubt there have been great improvements, and I must think mine the most perfect of all. Still, I feel great anxiety as to the results, for I will have not only the great disappointment and loss should it fail but also the "I told you so" of the whole country side.

April 11.

Began to mix the inoculating stuff for the alfalfa, boiling rain-water for the purpose. Elihu has ploughed with the heavy plough and Ball and Paul in the alfalfa field. Gibbie comes behind in the same furrow with Jack and Sambo and a bull tongue plough. They have gone very deep and the land should be in good fix after it. Made Willing try the Cahoon seeder to see if it worked according to directions on card.

April 12.

Elihu and Gibbie harrowing alfalfa field. I had a large tub on the piazza and put in the second ingredient for the wonder bath. I bought a corn planter this spring, not because I plant enough corn to really need it, but because the crooked planting of the women worries me so. To-day we were to plant the first acre of corn for this season. I had Willing use the planter drawn by Mollie. It worked very well, but he could not go straight and the rows look like snake tracks, much worse than the women's planting, and I had much better have saved my $10. Bonaparte is triumphant and I am in the slough of despond.

April 13.

Planted corn again. Had Elihu to run corn planter and had Willing to take his place harrowing in alfalfa field. The rows are a little straighter, but still hopelessly meandering. That $10 is simply thrown away.

April 14.

What a time I have had to-day. I started out to plant four acres of alfalfa and I feel just as though I had drawn the plough and the harrow as well as the three darkies. The land has been double ploughed, then harrowed with a home-made tooth harrow, and then with the acme several times. The land was heavily covered with stable manure before the ploughing. I have mixed the wonder bath most accurately and now the culmination of all, the planting, was to take place. I bought a Cahoon broadcast seeder, and have tried to make Willing (the boy I have in Jim's place, but oh, what a misfit!) understand the directions. I called upon old

Bonaparte this morning to measure the seed out into separate sacks, so that we would have no confusion in the field, but, oh, dear, what a dream that was! It seemed to Bonaparte such feminine folly that I should insist on stakes every ten feet at the head and end of the field so that Willing would have something to guide his wandering steps. We have had high words on the subject, he maintaining that it was a waste of labor and stakes to mark anything but the half acre. As Willing has not a straight eye and walks a good deal as though he were tipsy, even with the guiding stakes, I think it will be in the nature of a miracle if this field is covered with alfalfa. I have not been out here for two or three days, as I was planting corn, but I had two men and two teams at work all the time and a woman to clear away roots, etc., and positively I do not see what they have done. The field is as rough as possible, it seems to me, though the negroes think me most unreasonable and Elihu says: "My Lor', Miss, wha' yo' want mo'? Dis fiel' look too bu-ti-ful, 'e stan' same lik' a gya'ding!"

The first difficulty is to get the stakes set straight, a tall and then a short, so that Willing will know that when he leaves a short stake he must reach a short one at the end of the field; but I had a perfect battle to get Bonaparte to set the stakes in that way. The next trouble was to get rid of the alfalfa — I allowed ten quarts to the acre, and it will not go in. I have opened the small door of the conceited Cahoon creature just one-half inch as the card says, and made Willing walk every ten feet instead of every twenty, as it directs, and yet the peck of seed holds out and is left over.

I understand some of men's temptations in the way of speech now as I never did before.

Just here I am in trouble over the whereabouts of a huge caterpillar of varied and gorgeous colors which I saw a moment ago very near me. I did not like to shorten its little span of life, so I took it on a big leaf to quite a distance from where I was sitting and turned it on its back and made a little pen around it. Now it has disappeared and it may be anywhere. I must move to another tree, though I have an ideal seat on the root of this one, a splendid live oak with spreading branches.

Finding the ground still so rough I sent Elihu to "the street" to get a woman with a hoe to go over the ground and remove impediments. I said: "Get any one you can at once," thinking he would bring Snippy his wife, or Susan his daughter; but in a short time I saw a procession arriving. Aphrodite, with a basket on her head, a baby in one arm and a child of eighteen months dragging by the other hand, while one of three years toddled behind. The procession moved to a clump of trees in the middle of the field; there Aphrodite made a halt, took from her basket a quilt, and spreading it on the ground deposited the party upon it. I do wish I had my kodak; but I am so stupid about the films; I cannot put them in myself, and I am so afraid of spending an unnecessary cent, that for months my kodak is no use to me, and it would be such a delight if I could only once learn its intricacies.

Aphrodite spread a quilt and deposited the party upon it.

This group has saved my reason to-day, I think, for the little things are so funny, solemnly staring around, a bucket of rice and meat made into a strange mess in the midst. I sent for a basket of roast sweet potatoes, and gave one to each, but I disturbed the peace of the pastoral, for I insisted that the potato should be peeled for the baby, whereupon Isaiah set up a terrible yell and Aphrodite said: "Him lub de skin." I insisted, however, that the skin should be removed, for only a month ago Isaiah was at death's door with convulsions. The baby has on a little red frock and a little red cap with frills, tied tightly on her little coal black head, and the sun is broiling hot. Her name is Florella Elizabeth Angelina.

But back to the precious alfalfa, which has cost me so much worry as well as money All that I can get put into the land is six quarts to the acre. Here I pause with pleasure as another procession approaches. Oh, for my kodak again. I heard a noise, and on looking up I see the Imp puffed up with pride rolling the wheelbarrow, which seems to have a large and varied load. Behind comes my little maid Gerty with a basket. With a great swing Imp rolls the wheelbarrow alongside of me; and they proceed to unload. First a little green painted table, which has a history that perhaps some day I will have time to tell; then Gerty takes from her basket table-cloth and table napkins of snowy damask and all the implements and accompaniments of a modern lunch. Imp takes out a demijohn of artesian water, the cut glass salt cellar, pepper cruet, and then these are put in position and in the midst a little dish of butter, churned since I left the house this morning; and what a nice dinner! A fresh trout with a roe, brought me an hour ago as a present from Casa Bianca by Nat, broiled to a turn — a delicious morsel, and after that an abundant dish of asparagus, and besides this a large dish of fried bacon and one of rice.

"Oh, Gerty," I said. "Chloe knew I did not want all this to eat."

"Yes, ma'am," she answered. "An' Chloe say to tell you say we got plenty home for dinner en she know yu'd like to give some 'way." That made me happy, for Chloe to understand me so thoroughly, to send me a delicious dainty meal for myself, and then besides a substantial portion for me to give away. That is what an old time, before the war darky is, one whose devotion makes them enter into one's tastes and feelings so thoroughly.

When I left the house this morning I certainly expected to be back to dinner, but finding how absolutely necessary my presence in the field was I just stayed there, and at three Chloe sent this nice meal. When the procession arrived I exclaimed, "How delightful! Whose idea was the wheelbarrow?" The Imp answered promptly: "De me, ma'am," at which I made him my compliments. It is such a pleasure to be able to commend the poor little Imp, for he has an immense ingenuity in mischief and earns much reproof.

I am quite ashamed of the frame of mind in which I began this, but I will not tear it up. What is written is written. After this episode everything looks so different, and now at 4:30 the four acres are planted and 22-year-old Mollie is drawing a bush over to cover the seed with such rapidity that she keeps Elihu at a run, and even to my eye the field looks fairly respectable, and the darkies think it unspeakably fine. I am making Willing travel over between the tracks where he went before, and so have disposed of the necessary quantity of seed to within a peck. Now I can look up and beyond the gray earth and glory in the beauty of God's world. Half of the field was planted in oats in the winter and it is now splendid, an expanse of intense vivid color. The field, about twenty acres, is a slight elevation surrounded on three sides by a swamp, in which the variety of young green is wonderful. The cypress with its feathery fringe of pale grass green, the water oak with its tender yellow green, the hickory with its true pure green, and the maple with its gamut of pink up and down the scale — pale salmon, rose pink, then a brick-dusty pink, and here at last it rises into rich crimson. Here and there the poplar, with its flowerlike leaves, the black gum with its black tracery of downward turning branches, all edged with tender gray green.

It is too beautiful for words, and behind all, accenting and bringing out the light airy beauty, is the dark blue green of the solemn pine forest. I wish I had brought my crayons and block; I might have had a faint echo of one little corner to send to some poor shut-in who cannot get it first hand in its exquisite reality. And this, too, is but a prelude; in a few days the ideal tenderness will be replaced by a more material and lasting beauty, but not so heart reaching. It certainly seems a pity that one should have to think of and strive after filthy lucre in the midst of all this beauty; but I have reached a point where if I do not struggle and wrestle with the earth, therefrom to draw the said dross, I will have to give up all this life with Nature and find a small room in some city to eke out my days.

It is not a cheap thing to live in this country. One must have horses, one must have servants — but once given a moderate income to cover these things and there is no spot on earth where one can have so much for so little. Wild ducks abound all winter, also partridges, snipe, and woodcock; rabbits and squirrels run over everything. Our streams are filled with bream, Virginia perch and trout. If any one wants better living than these afford, he can have wild turkey and venison for the shooting, as the woods abound in these, and he can have shad daily during two months if he goes to the expense of a small shad net and a man to use it. It is a splendid country for poultry. Turkeys, ducks, and chickens are easily raised, and I believe it could be made to pay handsomely.

My first question to Gerty when she appeared to-day was, "How high is the incubator?" She answered promptly

101, by which I know it is not above 103, and am thankful. I fear the eggs are all cooked, for when I got in from the cornfield Thursday the mercury stood 106½. I had left Gerty to watch and to open the door if it went above 102½. She reads and writes and knows the figures quite well, but does not seem to understand the thermometer.

April 20.

I had told Aphrodite that she must pull up all the grass roots, brambles, etc., in the alfalfa field; as it was new ground the harrow had not got them all out. She came to me to-day and said: —

"Miss, I kyan't wuk een dat fiel' no mo'; de ting cum up too purty, en ef I tromple um I'll kill um."

"Do you mean the alfalfa has come up?"

"Yes, ma'am, de whol' fiel' kiver wid um."

I just flew to the field on my bicycle, and truly there was the whole field covered with tiny dark gray green leaves! I was perfectly delighted, for I had not supposed it would come so quickly and had no idea the stand could be so thick after all my tribulations.

Just before lunch S. came, bringing some friends with her — they wished to see how I turned the eggs in the incubator, and so I took the tray out to show them, and as I was putting it down on the table I heard a very soft chirp, which startled me so that I nearly dropped the whole thing.

Somehow I had not realized that the time was so near for the climax, but to-night as I was going to bed I went for a last look, and there was one little chick, white and fluffy and very lively. I wonder if that is to be the only one.

April 28.

The whole incubator seems to have turned into chickens. I never saw anything like it but a swarm of bees. As soon as I got up this morning I rushed down to the incubator, and there they were!

I called Chloe at once, and she stood in front of the glass door and gazed with wondering eyes, then she dropped a profound courtesy, and, raising her eyes and hands to heaven, she said, "T'ank de Laud," and this was repeated three times with intense fervor and reverence. Then she seized my hand and shook it violently.

Only then did I understand how much self-control Chloe had used not to show me more plainly her utter doubt and scorn of the 'cubator. I knew she did not approve, but had no idea that she felt certain we would never see a chicken from it. Her delight is unbounded.

The book of instructions says you must not open the door at all after the eggs begin to pip, but I had to open it very quickly and take out the egg-shells which were so much in the way of the chicks. It is too bad that they sent the brooder without any lamp, and so I cannot take the chicks out as I should do when they are twelve hours old.

The incubator must be kept at from 105 degrees, and the newly hatched chicks only 101 degrees, or at most 102, and so I am afraid of roasting the chicks or chilling the eggs.

April 29.

I am in a great quandary about the chickens, and I have to go to Gregory to meet a cousin at the train, for I cannot trust Willing to drive across the ferry and go to the station alone; he is too poor a driver, and so I must go myself. A great many eggs are pipped and the chicks will be sacrificed if I leave them so crowded and so hot.

After thinking it over I made up my mind, took a basket, opened the door of the incubator, took out thirty eggs which had not hatched, and going to the river threw them in. I stood on the little wooden landing and watched, and to my horror the eggs swam!

They would not go with the tide but made a circle and returned to the shore, and I felt like a murderer, but I could not get them back, so I sadly returned to the house and reduced the heat in the incubator to 102 and fed the chicks some bread crumbs. Then I got into the wagon and started for Gregory.

It was dark when we got to the ferry and I did not reach the Winyah Inn until 10 o'clock.

April 30.

When Willing drove to the inn for me this morning I saw a large red object protruding from his pocket, and as we drove to the station I asked him what it was. He appeared very much confused and would not answer, so I told him to take the thing out, as it looked very badly.

Finally with much difficulty I made him take it out before we reached the station, and it was a quart bottle of dispensary whiskey! I was very angry and told him to hand it to me, which he at first refused to do, but in the end he did, and I put it in my valise.

I told him I was greatly mortified and disappointed that this first time I had trusted him to drive me to town he should do such a thing. He protested and declared that it was for his grandfather. I was truly thankful I had seen it and disposed of it before M. arrived, for she had never been to this part of the world before and would have felt terrified to see the coachman so provided.

When we got home Willing's mother came and repeated the tale about the whiskey having been got for her father, and I gave her the bottle. I know this little tale is pure fiction, for her father never drinks, is a model old man, and I happen to know a piece of inside history about Willing, which he confided to Gerty, and she passed it on to Chloe, who in turn confided it to me, when warning me that my faith in Willing and his meek ways might be misplaced.

He told Gerty, who is his brother's fiancée, that he was "coa'tin'," but that when he went to see the object of his affection he couldn't say a word, but sat dumb before her, unless he drank a pint of dispensary on the way to her house.

"Then he could talk a-plenty."

Then he was all right and could talk a-plenty. I called for him this evening and gave him a serious talk.

I reminded him that when he was about five years old his father had gone to Gregory to pay his tax, having his pocket full of money from the sale of his crop. His poor mother walked the road all night with the baby in her arms hoping for his return. He was an excellent man, faithful to all his duties, a splendid worker, but he could not resist "fire water."

When I heard in the morning that he had not returned, and the other men who went with him had, I had Elihu get the pony carriage and drive down the road until he found him and bring him home, as the men said he had dropped asleep on the road and they could not rouse him, so they came on and left him. It was a bitter night, one of the three or four freezes we have during the winter, and I knew it would go hard with him.

Elihu found him eight miles away, got help and put him in the pony carriage, for Emanuel was a tall, heavy man, and drove rapidly home; but life was extinct when he reached the poor wife. I had sent beef tea and stimulant to be given him, but though Elihu found him alive, he could not force anything down; he seemed unable to swallow.

Lisbeth nearly went crazy; she had seven children to support by her own labors. As time passed she quieted down and having her house and firewood and two acres of land free of all rent and owning a fine pair of oxen and a cow, she got on very comfortably and brought up her children respectably.

When her only daughter, Aphrodite, married and her two oldest sons went to "town" to work and were making a dollar a day, she felt as though her troubles were over. But the same Devil's chain gripped and held her eldest son Zebedee.

He was a splendid boatman and was as much at home in the water as a duck. He owned a canoe and made an easy living, at the same time satisfying his love of sport by taking strangers out ducking. Many Northern people come to Gregory every winter for that sport.

Last January and February we had several bitter spells of weather with a prolonged freeze and snow. During one of these, when ducks were especially plentiful, Zeb took a stranger out. Late that afternoon they met another sportsman, paddled by a darky, and the parties spoke and commented on the unusual cold; and Zeb produced his bottle of dispensary, offering it to the other paddler, while his sportsman also produced a flask and urged it upon the second sportsman, who being near his home and its bright fire declined it and suggested to Sportsman No. 1 that he should land and not go on shooting, it was so cold.

No. 1, however, said he was all right, and pointing to his overcoat on the seat said he had not even put that on yet. They parted and Zeb and Sportsman No. 1 were never seen again alive.

They did not return to Gregory that night, nor the next. Then search was made, and the sportsman was found drowned and Zeb was found frozen holding on to some puncheons on the edge of an old canal. Near by was the boat, not capsized, and the things in it except the overcoat.

It was surmised by those who knew the circumstances that the sportsman, not being familiar with a dugout canoe, and not knowing that it is dangerous to stand up in one, rose to put on his overcoat, lost his balance and fell overboard, and Zeb plunged in to rescue him, a thing he could easily have accomplished under ordinary circumstances. But the spirits he had taken from time to time paralyzed his great strength and skill in the water, and he not only could not save the man but perished himself. He succeeded in reaching the puncheons on the edge of the canal, but was unable to pull himself out, and froze stiff there.

Of course I did not go into all these details to Willing, but made him see that without that fatal bottle Zeb could have saved himself and the man, and I tried to make him see that with such a family history the only hope for him was to swear off absolutely. He seemed much impressed and thanked me for my "chastisement," as they call any solemn counsel and admonition, and promised to heed it.

The chicks are very lively and eat bread crumbs and oatmeal very heartily. I have enclosed a space in the garden of fifty feet in circumference, with a netted wire fence six feet high, which I will keep locked, and I hope to defy hawks, foxes, and bipeds as well. Chloe is perfectly devoted to the chicks and feeds them with enthusiasm every two hours.

I am having much trouble at Casa Bianca. The hands

Chloe is devoted to the chicks — feeds them every two hours.

continue to resent my having given the keys to Nat, and they will not take orders from him. They will not bind themselves either to rent any certain amount of land, but sulk steadily.

I knew that the loss of my good foreman Marcus was irremediable, and when I met him in "town" the other day he told me he was perfectly wretched; that he missed the country so. Of course it must be so at first.

Instead of using his really excellent powers of control and organization, he is hauling wood for a living during the week and preaching on Sunday; but his wife is perfectly happy in the high social life. It is the old, old tragedy of Eve and her misguided ambitions — the world, the flesh, and his satanic majesty. The apple pleased her eye; she longed to taste it, and then the subtle whisper came: "And it will make thee wise."

Marcus was making a handsome income; had a position of trust and responsibility, where all his faculties were in use during the week; and on Sundays he, no doubt, preached good, simple, useful sermons to his congregation of laborers, for he came fresh from his struggle with the earth and its realities. But to his wife came that desire for social eminence; to wear silk frock and shine, and she tugged and tugged until he consented to her going.

He remained a year alone on the plantation and then came the inevitable. He followed, and now all the dignity of his life and character has gone, and he is struggling to make himself contented with what is supposed to be a higher station; that is, he takes orders from no one. He will get accustomed to it after a time, but his powers will shrink away, unused, and without responsibility his character will crumble.

When he began as my foreman,* about fifteen years ago, his

---

* Marcus has since died. He was found one morning in his stable, where he had gone to harness his horse, leaning against the manger, stiff in death. He bore a high character, and his death was regretted by white and black.

writing was illegible, his figures hopeless. Steadily, patiently, I have corrected his mistakes, looking over and deciphering his weekly accounts and copying them down in my book before him so that he could see how they should look. Now he writes a readable, nice letter and any one could examine his accounts, and he knows and realizes all this and knows that his standards have all grown and risen more even than his knowledge.

Meantime I will have to give up altogether planting on wages, and it looks as though there will be very little land rented. If I had money of my own I would hire a good overseer and plant 100 acres on wages and not rent any land to these recalcitrant hands, but it would be madness to put a mortgage on the place and borrow money at 8 per cent while rice is selling at 40 cents a bushel.

So I will simply remain passive and let the hands who wish to rent have the land and seed, but explain that I cannot pay out any money for extra work. I feel sure that some day rice will rise in price, but every one seems to think differently, and all the planters are either giving up entirely or diminishing their acreage very much and turning to upland crops.

So far I have only forty acres of rice land rented, and I feel very blue about the future. Then, again, my sheep and cattle at Casa Bianca, which have been so remunerative to me all these years, are giving me trouble now.

A friend and neighbor, who has been heretofore a confirmed rice planter, and never planted an acre of corn, has become disgusted with rice and enclosed a large body of land which has been thrown out for years, and is going to plant corn and cotton. This land touches mine, and my animals have had the run of it. The fence which has been put up is neither "horse high, bull strong, nor pig tight," and my cattle do not regard it at all, though it is a very nice looking, *comme il faut* wire fence, and I will have to sell my cattle, I fear, and confine the sheep in a limited pasture.

Ruth, my brag cow, who has given me fifteen fine calves, and Rubin, my picture bull, just light over that neat fence as though it did not exist, and the humble sheep go down on their knees and creep under it, and I lie awake at night and wonder what I am to do between my love for my creatures and my love for my neighbor.

CHAPTER V

Easter Sunday, May 1.

A BEAUTIFUL, bright Easter. All nature seems to rejoice with man in this great day of triumph over death.

Our little chapel, Prince Frederick's Pee Dee, is beautifully wreathed with wild flowers and vines, the work of three

Prince Frederick's Pee Dee.

young girls, sisters, who, having but three days' holiday from their school teaching, devoted one of them to this thank offering and labor of love. We are all touched and softened by this act of devotion, and the blessing of the day seems upon every one.

May 2.

Had a terrible shock to-day. I took M. to see the alfalfa field, and there was not a leaf of anything in the five acres! Those two nights of ice must have caught the alfalfa in its one tender stage, for all the books say that after it is six inches high it will stand any amount of cold. I am stunned, it is such an unexpected blow.

Having been desperately busy, and knowing that my fence was perfectly secure, I have not been to look at the alfalfa since the seventeenth, when it was fine, and now all the money I have spent on it might as well have been thrown away, so far as any hope of return goes — I fenced in that field of thirty acres with American fence wire, forty inches high, and two strands of barbed wire on top, hoping gradually to get it all in alfalfa by planting five acres every year. I have five acres of fine oats in it now, but that brings in no money, only feeds my horses.

I had to go for a long walk alone to steady myself, so as not to break down entirely.

CHEROKEE, May 3.

The hands from Casa Bianca came this morning to get seed rice. I was just starting to drive M. to the train, but as it is very important to get the rice planted as soon as possible I had to delay the departure until to-morrow, for it was too late when I had finished measuring out the rice to drive to Gregory in time for the 4 : 30 train.

May 4.

Drove M. to the R. R. yesterday. I was afraid to take Willing, knowing his weakness for the dispensary; so drove her in the buckboard. On the way I took her into Woodstock, my brother's place, that she might see its beauty, and then when we reached Gregory I took her to see the old church, Prince George Winyah, and its churchyard, where my parents rest. The church was built of brick imported from the old country, and it is one of the oldest in the land. The churchyard is beautiful with its moss hung oaks and cedars, and one feels that it is truly God's acre. We lingered there so long that there was a risk of missing the train, which would have been most inconvenient to both guest and hostess. By driving rapidly, however, we reached the station in time.

As it was too late for me to take the long drive home alone I went into Woodstock and spent the night with my brother.

Prince George Winyah.

This morning after breakfast I drove to Casa Bianca, which is halfway between Woodstock and Cherokee. There I had a good many things to see after, and it was late afternoon before I got through and finally started for home.

I had been so much engrossed with my work trying to establish a better state of feeling between the hands and Nat that I had not noticed that the clouds had gathered heavily and that everything indicated a storm. When I felt the gusts of wind which tore at the umbrella so fiercely that I had to put it down in spite of a drizzling rain, and saw the forked lightning which shot incessantly from the clouds, and thought of the eight miles of lonely road ahead of me, I realized that I would have to bring forward all my faith and philosophy for the next hour. From being by nature a great coward I had become very courageous, and I have often caught myself saying there were only two things in the world I was afraid of, a cow and a drunken man, and I could not help calling this to mind now and wondering how I would stand the present ordeal. Romola, who is generally very quiet, snorted and showed every sign of fear, but I did not give her time to give way to her feelings, but used the whip freely, a thing I very rarely do, to make her understand that she must travel. She responded nobly and we sped along.

The clouds made it much darker than it should have been, for the sun had only just gone down. I have never seen such vivid lightning nor heard such claps of thunder, and at each Romola darted out of the road as though the thick bushes could protect her. Not a human being was to be seen the whole way, and when I got to the avenue gate, which was shut, I had, of course, to get out to open it, and I felt sure Romola would fly home and leave me; but I did her an injustice. She waited, with every sign of impatience, long enough for me with great speed to get in, and then dashed on until we got to the darkest spot in the avenue, where the live oaks lap together overhead. A fearful flash of lightning came, followed instantly by a terrific peal of thunder, and she stopped short. I felt sure she had been struck, and she seemed to share the impression, but in a moment she went on and we were soon at home.

I was so excited that I was in a perfect gale of spirits,

which quite upset my good Chloe, who had worked herself up to a wretched state of anxiety about me, miserable that I was out in that terrible storm alone; and she was hurt and disapproving of my attitude, especially as the first thing I did was to insist that Gerty and herself should take in my best rug, which had been hung on the piazza to air. Their terror had been so great that they had left it out in the rain — such a panic had seized them that they were very reluctant to venture out on the piazza. They had the house shut up without a breath of air, that being their idea of safety. Of course, I was drenched and had to change all my things, and after two hours I sent word to Willing that he might safely feed the mare, I having told him to rub her perfectly dry, but not to feed her till I sent him word. What was my dismay to find he had not rubbed her at all — said he was afraid to stay in the stable, so he had turned her loose in the stable yard and gone into the kitchen, leaving her exposed to the pouring rain! Of course she will be foundered, for she was very hot.

Sunday, May 8.

Drove Ruth to church and met some one just from Gregory on the way, who told me a most terrible thing. Mrs. R., one of the loveliest women in our community, was struck by lightning during the storm last evening. She had always had a great terror of lightning, though in every other respect she was a fearless woman, so that her family always gathered round her during a storm and tried as much as possible to shut out the sight and sound. On this occasion her husband and daughter were sitting one on each side of her on an old-fashioned mahogany sofa, she with her handkerchief thrown over her face. When the fatal flash came the husband and daughter were thrown forward to the floor and were stunned; as soon as they recovered consciousness they turned to reassure the mother as to their not being seriously hurt. She was still sitting straight up on the sofa with the handkerchief over her face; they lifted the handkerchief as they received no answer and found life extinct. It was a translation really for her, as she probably felt nothing; there was only one small spot at the back of the neck. She was a woman rarely gifted, with beauty of face and form, as well as of soul; she was one upon whom every one rested who came in contact with her; she gave of her strength to all who needed it, for her supply was unlimited, coming direct from the great source of all power. I wonder if terror of lightning was a premonition which had been with her always from her childhood? Her death is a great loss to our county, and to her family a calamity indeed.

May 9.

Very busy arranging things so that I can leave for my annual visit to Washington. It is harder than ever, for Jim not being here to leave in charge of the horses I feel very anxious. However, I have done my best and will leave tomorrow. The incubator is in full swing and Chloe and Gerty have learned how to manage the heat between them. The chicks are due to hatch on the 14th, and I have left most accurate written directions for each day which Gerty is to read aloud to Chloe as the day comes, for toward the end the heat must be raised. The first family of sixty-seven are growing apace; only one has died and that was smothered by the others before I found out that I must put them under the hover every night or they will cluster about the thermometer and climb on top of each other until the ones underneath are smothered if help does not come. It is the funniest thing to see their devotion to the thermometer. They peck it off of the nail on which it hangs, so that as soon as I learned to know the proper heat for the brooder by touching the metal cylinder under the hover, I took the thermometer out entirely, and as soon as it was gone they went under the hover of their own accord. They seemed to feel that the mercury was a living presence, I suppose, because it moved up and down in the tube.

I am leaving Willing to run one cultivator, with Mollie and Gibbie to run the other with a fine ox I have just bought. I heard that Gibbie had made his plans to "go to town" to work, leaving his young wife and child, and I racked my brain for something that would interest him at home and divert his thoughts from that plan; for if once a young negro leaves his wife and children to go away to work he is very apt to stay away permanently, and I should be sorry for Gibbie to do that. One day I called him and said: "Gibbie, I wish to try an experiment and put you in charge of it, and I am going away for a month. You know, in this country no one ever thinks of ploughing a single ox; they can't do anything without a yoke of oxen; but in the up country it is not so. On my way to the mountains I see from the car windows people running their ploughs with a single ox. Now I want you to take entire charge of Paul — no one else is to use him — and I want you to put him in the cultivator and run it through the corn day by day until you finish that, and then through the cotton, and then start through the corn again; but be careful of Paul and do not let him get galled, and feed him well."

Gibbie was as proud as though he had been made Viceroy of India and his plan of deserting vanished.

May 26.

Washington. Spen the afternoon at the Agricultural Department, where I met with much courtesy as well as information. I went specially to inquire as to the practicability of the cultivation of the orris root on our rice field lands. The orris of commerce is the root of the iris, which grows luxuriantly in our low country. In the latter part of March and during the month of April every swampy low spot, as one drives along the road, is beautiful with the dark purple or blue and the light purple and the white iris, or flag. My desire was to find out if these species of iris had the perfumed root, for if they have we could cultivate it in the rice fields with great success.

The impression at the department is that orris can be grown only on high ground, as in Italy, where it is principally grown, it is planted in a semi-mountainous region. This is a great disappointment. They told me of a farm in Louisa, Va., where the orr s is being cultivated for market. I would like very much to visit that farm and see for myself, but my time is limited, as I have promised to attend the annual meeting of the South Carolina branch of the Women's Auxiliary at Orangeburg, May 31. One must have plenty of patience to attempt the cultivation of orris, for the root should not be dug until it is two years old, and then it has to be kept two years before its perfume develops.

Another thing I had much at heart was to take some lessons in photography and to buy a good camera. I could do so much more if I could illustrate things with good photographs of the odd and picturesque things I so constantly see; but, alas, I am going away without having made any progress in this direction, time and other things lacking.

June 6.

Peaceville. At home once more and the great big white rooms of the pineland bungalow are very restful and pleasant. That is the one luxury we enjoy to the fullest in the South — space. My rooms here are immense, each with four windows and three doors, very high ceilings and a broad piazza around the whole.

I received a riotous welcome from the dogs and a very hearty one from Chloe, Gerty, and the Imp, but Chloe seemed downcast and unlike herself, and I knew there was some bad news, which she would not bring out until I had had my dinner. While I was away I had several letters from Chloe, in one of which she announced with great joy that sixty-three fine healthy chicks had hatched from the 'cubator. So when I had finished the simple but delicious meal which she had prepared for me I asked her to go out with me and show me the chickens. Then she poured out her woes. The night before she moved from the plantation some one had climbed the six-foot fence and stolen twenty-five of the precious last-hatched chicks. She said when she found it out the next morning she sat down and cried, she had been so proud to have hatched them out and they were doing so well and growing so fast. I sympathized with her. Of course it was a great blow to me, but she was in such deep distress over it that I had to act the part of consoler, though I was the victim.

She went on to say: "En I do' kno' who carry de news out say I cry 'bout de chicken, but I s'pose 'twas dat wicket boy Rab, fu' ebeybody I meet say 'Eh, eh! I yere say yu cry 'bout chicken, I'se shock to yere sech a ting! A pusson cry fu' loss 'e mudder or some of 'e fambly, but cry fu' chicken! No; en wusser wen 'tain't yo' chicken.'" This taunt and ridicule seemed to have sunk deep and to rankle still. She went on to say that the person who took the chickens must have been well known to the dogs, as they made no outcry, and moreover that Rab had not slept at home that night, saying he had stayed with Willing, which all looks very bad for both of these boys. I will not attempt to investigate, for it would be perfectly useless.

It is a principle firmly maintained that one negro will not give testimony against another unless he has a quarrel with him, and then he will say anything necessary to convict him of any crime, so that investigation with a view to justice is a farce. I do not doubt that these two are guilty, for Willing has encouraged Rab to return to his old habit of stealing all the eggs. Bonaparte found a spot in the pasture, with cans and many egg-shells and remains of fire, where they had a regular picnic place. When he asked Rab about it, he said Willing and he cooked there every day eggs, potatoes, any-

"Eh, eh, I yere say yu cry 'bout chicken."

thing else they wanted. I had brought Rab a beautiful outfit from Washington, besides the ever desired mouth organ, and, after a consultation with Chloe, I determined to give them to him, as she said he had been moderately good while I was gone and slept out only that one night; and there was no proof against him, and if they did take the chickens of course the older boy was very much more to blame. I would not on any account accuse them of such a thing unless I was perfectly certain, for I think that is the way to make people dishonest. I would not appear to think it possible that any one about the yard could know anything about it. I only reproached Rab with having been absent that night, as he might have caught the thief.

June 20.

Drove into Cherokee this morning on my way to Casa Bianca and found Ruth with a beautiful filly colt. I am so pleased. Ruth is very proud and brought the colt right up to me and the little thing licked my hand and let me stroke its head. I went on in fine spirits after admiring my new possession. So many things go wrong that I am unduly elated when something pleasant comes.

The Summer Kitchen at Cherokee.

Casa Bianca looked perfectly beautiful. The place is so lovely that it always does me good to go there, though this time I had dreaded it very much. The negroes continue to fight against Nat, and there is very little rice planted, and they will not work that little properly. Nat seems to do his best, which I'm sure is a mercy.

Stopped on my way back and told Willing to get all the milk he could to-morrow and put it in demijohns ready for me to take out with me. We are to have a little sale of ice-cream in aid of our auxiliary.

June 21.

Arose at 6 and hurried through breakfast to go early to the plantation and get through my work there and bring the milk for the auxiliary. To my great disappointment found Willing had less milk than usual. I went with him to the field and made him milk the cows over, and found they had an abundance and he had only half milked; he was sulky about it, but I insisted and got three quarts more, then turned the calves in and showed Willing how much they were getting. I hurried back to send it to the ladies. I had undertaken to furnish the 200 pounds of ice and to make a churn of cream myself. Such a time as I had freezing it! I never had done it before, as long ago I read all the directions to Jim, who always did it. I supposed Chloe knew how, but she had forgotten, if she ever knew, and I spent nearly two hours down on my knees working with the thing. Like everything else, it is easy if you know how, but this was terrific. However, it was finished in time.

The Winter Kitchen at Cherokee.

In the little hamlet of Peaceville truly the simple life, now so much vaunted and preached, is lived. A community of gentle folk, about sixteen households, most of the families were wealthy in the time prior to 1860, and all well born. Now theirs is a life of privation and labor, and borne without murmur or repining, and they are gentle folk emphatically. With the mercury for weeks over 90, and sometimes 98, there is but one family who can indulge in the luxury of ice. Until this summer I have always got 200 pounds a week, but things are changed by the failure of rice and I have given it up, as by the time I get it from the nearest town, eighteen miles away, the 200 pounds cost $1.50. Every one is much excited over the sale, and early in the afternoon they gather at the little schoolhouse, across the road from my gate, which had been selected for the event. The five ice-cream churns are grouped under a tree and two or three tables placed around, while the benches from the schoolhouse are placed about as seats. Two ladies down on their knees serve out the cream to the excited string of children, who bring their nickels clasped

The string of excited children.

tightly in their hand. Two other ladies have a large dish pan and towels and keep a constant supply of fresh saucers and spoons, while one with a little basket receives the nickels.

The five-cent saucers are very big, but I call to mind how rarely these children ever taste ice-cream and what self-denial on the part of the mother each nickel represents, and so our results are not as large as they might be. My churn is pure cream as that is the only kind I can make, but it is not nearly so popular as the others which are made of custard with different flavorings. Finally, after a period of great activity I hear "All gone but the Newport vanilla" (that is mine) and the answer comes, "Well, if there is nothing else I will take that," and everything is gone and the benches are put back in the schoolhouse and the tables are carried home, and we have made $8 for our auxiliary, not much, but it represents a good deal of labor and self sacrifice on the part of the women who have given their material and their time, for al the things are contributed by different members and so we have no bills to pay. This will go to a cot in the hospital at Shanghai in memory of Bishop Howe and for a Bible woman in Japan. A mite truly, but God grant it may be blessed.

June 22.

Rose at 5, skimmed and set the churn. It is very hot, and having no ice there is no chance of good butter except by handling it very early. When I went to the plantation, I found that my two English side-saddles had been left on a rack in the piazza where I had them moved this spring from the stuffy harness room, but I didn't mean them to stay there always; it is scarcely safe now that I have moved to the village and there is no one in the yard; so this afternoon I called Gibbie to bring them into the house. He brought the first and placed it on the rack, and I covered it with a large white cloth and he went for the other.

As he came with it I heard a strange rumbling noise. "What is that?" I asked. Gibbie is quite deaf and answered that he heard nothing. I went on: "It is either a steamboat on the river or an approaching tornado."

Still Gibbie heard nothing, but as he was about to put down the saddle I became aware that the noise came from it. "Take it back quickly to the piazza, Gibbie, and put it down gently." I followed, and as he set it down out from the inside crawled a bumblebee, and then another and another. The bees had excavated the padding and built inside of the saddle, leaving only the small hole which they had bored visible. The saddle might have been put on a horse's back and girthed on before the bees stirred, and what a circus there would have been.

Nothing would induce Gibbie to touch it again. He fled down the piazza steps, and the saddle remains upside down on a stand. I do not know how to get rid of the things. The sting of the bumblebee is said to be more severe than that of the honey-bee. If I pour hot water down the hole, as I first thought of doing, the saddle will be ruined, and I do not know how else to reach them. Certainly strange things happen to me!

When I reached Peaceville at three o'clock the mercury in the coolest spot on the piazza marked 96, and I was so thirsty. Alas the artesian water brought and kept in a demijohn is lukewarm and there is no use pretending that it is refreshing. The well water is cool, but it has a taste which makes me prefer the tepid contents of the demijohn. I have made great efforts to cool it; sewed it up in cloth and swung it from a nail in the piazza — all in vain. From contrast to my expectations, I suppose, it seems hotter than ever, so I gulp down the clear liquid, saying to myself, " you are obeying one of the first laws of health in not drinking cold water — only fools fill their digestive organs with icy fluid."

PEACEVILLE, July 18.

Rose at 5 o'clock and had breakfast before 6 o'clock, so as to make an early start to drive to Gregory to attend the farmers' meeting and hear the lectures by the agricultural experts. The heat is so great that the early morning or late afternoon is the only time to travel.

I got through the eighteen mile drive very quickly it seemed to me, for it had looked interminable to my mind's eye when I started, and had an hour in town before the meeting.

The hall was quite full and I was very glad when D. came forward and met me and ushered me in. It was quite a tribute to him that so many of the most primitive farmers came. They looked quite lost until he met them at the door and found seats for them with a delightful courtesy and interest.

I saw many that I knew rarely left the deep recesses of the pine forest, and it was quite touching to see the attention with which they listened. Near me were two I knew well, quite young lads, whose life had been spent in a struggle with the soil, beginning as small boys; Colonel Ben and Solomon.

I had handed the former to the Bishop to be christened. His father had selected as his name "Colonel Ben." Fortunately I asked the Bishop about it before the service began, and he answered, "He cannot be christened Colonel Ben. They can call him by that name, but the title must be left off in the service."

When I repeated this to the mother she was very stolid and said, "Par, he named him Colonel Ben, en he wishes him baptized the same."

I understood it entirely. They named him after the man who had done most for them in their lives. He had been a Colonel in the Confederate army, and after the war became a clergyman of the Episcopal Church, and their desire was to name the baby after the Colonel and not after the priest.

Poor Colonel Ben has had a hard, limited life, but has worked faithfully tilling the soil on his father's large farm. The unwonted excitement of a visit to the county-seat to hear a man tell him how to do what he'd been doing all his life was most astounding. As they tiptoed behind D. into the rather dark room filled with people I think it would have taken little to make them turn and run to the shelter of the woods.

However, they settled down and after furtive looks around devoted themselves to trying to make out what the speaker was saying. For a long time it seemed to me they were getting nothing. It was all a confused talk to them, and then he said something which roused them to interest: "And now I will tell you how to get the greatest amount of good from your barn-yard manure," and he proceeded to urge them to haul it on to the fields as fast as it accumulated.

Both Colonel Ben and Solomon leaned further and further forward in their desire not to lose one of the precious words of wisdom. It was lucky that the two seats in front of them were vacant, for the long arms were far over the seat, while the eager faces tried to bring the huge hearing members nearer to the speaker. I felt quite delighted that they had found something available, something they could carry home.

It is hard for an educated speaker to realize how his fluent speech slips off the rustic brain like water flowing over a rock. They cannot absorb it; it is all over before they have caught on.

After it was all over I met Colonel Ben, Solomon, and their father wandering along the street. I stopped and spoke, asked them if they were going to the banquet which had been prepared for the audience.

No, they reckoned they'd be gittin' on home.

But I urged them, saying I felt sure R. L. A. expected them and would be looking for them.

"Wall, he's the one got me en the boys into this trouble; he wouldn't take no, we jest had to cum, en hare we is."

I started them on the way to the hall and hope they got there and enjoyed the substantial lunch provided. No doubt these meetings do an immense deal of good if as in this instance the local director is a man of enthusiasm and able to throw it into the work and take an interest in all the individual farmers who are so cut off from the interests of the rest of the world.

They think that to scratch over many acres of land, guiltless of manure or help of any kind, with a yoke of oxen and then to have all the family from the oldest to the youngest turn out and plant the corn by hand, disturbing it as little as possible by work until it is ready to harvest, is to be a farmer, and they are satisfied. In the spring R. L. A. was trying to persuade one of these very satisfied old men to plant a few acres under the direction of the Department. He turned on him.

"Look a' yere, young man," he said, "I bin fa'ming long before y'u ever wus thought of, en I want y'u to onderstan' I don't believe in deep ploughin', I don't."

R. L. A. used all his blandishments until the old man promised to plant two acres by his directions, beginning with deep ploughing. He told me that when he went back some months later the old man said: —

"Youngster, I don't know what's the reason, but I kyan't get any of my corn to grow but them two akers o' yourn — the dry drought is just a-burning up the rest o' my corn."

And still later when the steady rains set in and he went that way the old man clapped him on the back and said with much embellishment of action: —

"Well, you've got me; the rain's done finished the rest of the corn, but them akers of yourn jest keep on a-growin' en a-growin', en I jest tell you now next year I plants jest about half o' what I bin a-plantin' en I ploughs it all deep en does jest es you tells me to do."

That was a wonderful triumph for the young director, and he tells me there are many such cases.

July 21.

Having land prepared for turnips, which are a very important winter crop for us. The corn and cotton both look very well, also the potatoes, and the little amount of rice planted is fine. The agricultural society of the State has offered a prize of $100 for the best results in hay from five acres of alfalfa during 1906, and I have determined to enter the contest. I know I cannot get the prize, but trying for it will make me more careful in planting and preparing the land. They give very exact directions and insist on a great deal of fertilizer being used — that is, what seems to me a great deal, and I never would spend all that money unless I were in a way forced to it by entering the contest. I am now reading everything I can find on the subject of alfalfa, and there is a great deal to be found.

Wrote to George T. Moore for inoculating material for alfalfa. I am so delighted that he is back in the Department of Agriculture, so that I can write to him. I have been miserable over what I considered the great injustice to him, and am so thankful that amends have been made and he has been reinstated.

I am so happy to-day over a check received from a liberal paymaster that I am quite stupid. I had sent off the last money I had in the bank for fertilizer for the alfalfa, and was feeling anxious, and now I am so relieved.

CHAPTER VI

PEACEVILLE, July 23.

WITH great difficulty got Chloe off to Gregory to make a visit to her daughter and see her grandchildren.

I have to push and force Chloe to take the smallest holiday or relaxation. She cannot drive, so of course I had to send Dab to be her charioteer.

I told her to broil a nice, 'cubator chicken and put it in the safe, and I have a very nice loaf of bread which I made yesterday, and with delicious fresh butter and tomatoes I will be independent of cooks for two whole days.

In this blessed hamlet of Peaceville, Bible methods prevail to a great extent, and people do as they would be done by. One finds out what vegetable one's neighbor is short of, and if you happen to have that special thing in abundance, you fill a basket, put a dainty doily over it, and despatch your inevitable small boy to your neighbor with a pleasant message. Of course she is too delicate to return her abundant vegetable by the same messenger, but later in the day or the next morning arrives her small boy with a dainty covered tray, and you receive a supply of the vegetable you lack with an elegant note.

I planted a great many tomatoes, but for want of work during my month's absence they are very backward, while my dear friend and neighbor, in the best sense of the word, Miss Penelope, has an abundance of large, smooth red tomatoes, and daily I receive a little tray of them. I have only very prosaic vegetables as yet, beans and Irish potatoes, but

198   A WOMAN RICE PLANTER

I got Chloe off to make a visit to her daughter.

they are fine and plentiful, so Dab makes expeditions with the tray, but without the note which is Peaceville etiquette — a little note asking particularly the state of your health and mentioning the height of the mercury, and saying that though doubtless it is sending coals to Newcastle you venture to offer some of your poor products.

I have a box tacked on the wall by my writing-table into which I drop all the notes received. I keep them, for they breathe such kindliness, and seem an echo of the past when people had time to think of others. By the end of the summer they would nearly fill a half bushel.

To-day I tried to conceal the fact of Chloe's absence. I was invited out to dinner, but I was so exhausted after the service that I was not equal to going. Though I had made every effort to get Chloe off before service, she was not ready when I left, so I told her to lock up the house and put the key under the pot of heliotrope on the shelf in the piazza, where I found it, and opening the door, which gave light enough for me to read by, I lay on the lounge in the dark, shut-up house till afternoon, when I felt sufficiently rested to get up and take my frugal, but delicious, repast of cold chicken, bread and butter and raw tomatoes. Thanks to one of my unknown, far-away friends, I can enjoy my glass of artesian water. He wrote from Saratoga, suggesting that I should fill a stone jug with the artesian water, attach a long rope, and sink it in the well. I have done this, and by this simple expedient I have delightful, pure drinking water at a temperature of 63 degrees, without having ice. When the mercury is soaring in the 90s 63 seems cold, and I do not ask for better. Except for keeping the milk and butter and having a treat of ice-cream occasionally I really do not miss ice, now that my little brown jug is swung in the well, and I am very grateful to my far-away friend.

At dark arrives Miss Penelope bearing a large tray, "Oh, my dear, I have just heard that Chloe and Dab were seen this morning driving out of the village and have never been seen to return! And to think of your being all alone here, and we not knowing! And we had such a delightful dinner! If only I had known! But I have brought you a bowl of cold okra soup and a little dish of ice-cream, for we had a celebration to-day, a birthday dinner."

As soon as I could I told her I had had a dinner fit for a king, and now this wonderful and delicious treat of ice-cream made it perfect.

I really do not miss ice, now that my little brown jug is swung in the well.

I read S. D. Gordon's "Quiet Talks on Power" all day in the darkened room, and I feel as though I might develop into a dynamo of the first order.

Peaceville is one of the corners of the world where Sunday is carefully observed. No one thinks of reading a novel or even a magazine on the day of rest. The *Spirit of Missions*, the *Churchman*, the *Diocese*, and sermons are the only mental food digestible on that day. I often find myself reading "The Spectator" in the *Outlook*, but if a neighbor comes in I put it hastily out of sight.

Last Sunday my dear Miss Penelope, whose whole life is a sermon of unselfish devotion and service, never resting, for on Sunday after her laborious six days in the "store" — which has done such wonderful things, supporting the family, educating all the younger members and finally paying off the mortgage on the plantation — she is our sole dependence as an organist and she never fails us. She understands the "instrument," as the organ is always called by Peacevillians.

This special organ is her own, which she has lent the church. The one which belongs to the church was originally a fine organ and was given to this little chapel some thirty years ago by a rich young man in New York, who had it for his own use and who was dying and expressed the wish that it should go to some place where it would do good. Our little church was without one and the rector happening to be in New York at the time, it was given to him. From that day to this it has been in use constantly, and without repair.

Church mice are proverbially active and they showed great fondness for the material of the bellows, so that the "instrument" was in a sad and wheezing condition, making respectable sounds difficult. I, being an optimist of the first water and having received constant proof of my having the right view of life, said boldly two years ago, just before the storm which laid us all low, that I would undertake to pack the organ and send it away to be repaired. A tuner who came to tune my beloved piano that spring said that he would repair the organ for $40 if we sent to it him in Carrolton.

I had had experience of the great liberality of the makers of my piano in the matter of exchange. During a period of forty years more or less I have had four pianos. Each time that by some good fortune I felt I might give myself the blessing of a new piano I wrote to the makers and told them I was sending on the old piano and wished a new one. They always allowed me a handsome price for the old piano. Reasoning from this experience that all great makers would act in the same way, I wrote to Boston to say I was sending the organ for repair or exchange, as seemed best to them, and asking their best terms, stating that by this parish there had been bought five melodeons of their make, including two baby organs.

Immediately came a letter to say the repairs would cost $80 and when they were made the organ would be worth $250. I wrote back in despair to say we could not possibly raise $80 for repairs, but would accept any melodeon they would send in exchange for our organ, which by their estimate was now worth $170. The answer came promptly — they could not offer any exchange; the organ was in their way; please answer at once.

So here am I, having sent off the property of the church on my own responsibility, and it will probably lapse by dint of possession before we can possibly raise $80 for the repairs. At night when I am very tired, the organ has a way of rising up before me in accusation and I feel it is an "instrument" worthy of the Inquisition.

It has been two years now an unwelcome guest in its childhood's Boston home. Meantime we are using Miss Penelope's organ, which is not fair, for she can never practise the hymns at home, having no instrument.

I began all this to tell of Miss Penelope's temptation. Last Sunday afternoon the unwonted sight of an automobile struck the village. Great excitement among all those who were so fortunate as to be strolling along the dusty road, among whom was Miss Penelope.

The occupants proved to be friends of hers and when they got out to make a visit in the village they asked her and two ladies with her to get in the machine and take a little turn. Now, Miss Penelope had never been in an auto and she accepted at once. They went two or three blissful miles and then came the awakening. Every face they met was set in solemn wonder that she, Miss Penelope, a pillar of the church (if the church is ever allowed confessedly to rest on feminine foundation), should ride in an auto on Sunday.

Words failed, but looks were all powerful. That night she said to me: —

"Patience, I am so ashamed of myself; I just yielded to temptation, you may say, without a struggle. It was so hot and dusty in the roads and the thought of flying through the air was so delightful that I never thought of it being Sunday and accepted the invitation at once; and it was the most heavenly sensation! Mr. A. said the road was clear and he could exceed the speed limit without danger, and it really was like a trip to Europe, so elevating and delightful; but as soon as I stepped down from the car I realized how wicked I had been."

"My dear, I do not agree with you at all," I replied; "there were no horses being driven for pleasure on their day of rest; there was nothing but the cogs and wheels of a machine and half a pint of gasolene. You were perfectly right to go. Don't mind what any one may say. It was a perfectly innocent recreation and refreshment, which you of all people are certainly entitled to."

But my efforts were in vain, though she said: "It is a great comfort to find you do not blame me, but I must blame myself."

July 24.

Good Miss E. spent last night here so that I should have some one near, and she made me a delicious cup of coffee and a nice little breakfast in spite of all I could say. Then she went home, and I fed the chickens and washed up the dishes and did all the housework, drawing buckets and buckets of water from the well, and I felt so proud and pleased with myself when I found it was only 9:30 and I had done all the work, for I had to do Dab's as well as Chloe's.

It is a great thing to know just what the work is, and if you do it once yourself you know just what the labor is. It is not a third of the amount of work I had supposed.

After finishing I sat down in the door of the sitting room to get every breath of air and embroidered and had a day of luxury — no interruptions, except when one waiter arrived with tomatoes, another with a muskmelon, and just at noon a specially dainty little tray with a glass of blackberry shrub and, O joy, a lump of ice in it.

I do not know when I have had such a quiet, peaceful day. As the horse and vehicle were gone I had no way of going to the plantation, which is my daily duty, and so felt free to enjoy myself.

July 28.

My poor Chloe is very ill with rheumatism — it is distressing, she suffers so. Dab is distinguishing himself and so am I.

I rise at five, so as to churn and knead and do my part of Chloe's work. Dab does the cooking very well and with enthusiasm. I am conscious that with both of us it is the enthusiasm of new brooms and am looking with terror to the inevitable slump.

I have never been an early riser, and the thought of the stern resolution I have made, to get up at five punctually, keeps me waking up all night long. I strike a match, look at the little clock on the table at the head of the bed, and think with delight how many hours there are before the fateful five strikes. I am losing pounds daily in the process, but make up in pride over my strenuosity.

On the plantation the struggle to get all the peas ploughed in for hay is most exhausting. Gibbie says he is sick, and I have engaged Loppy to do it, but he finds fault with the team and the plough.

Sunday, July 30.

A pleasant service and good sermon in our little church. When I got home to my great joy found C. and John. We got out the little old black leather-covered trunks which came from the log house, where they were stored, and looked over some of the papers in them.

Found many old letters from grandmother to papa when he was at West Point, beautiful letters, urging on him duty, discipline, and diligence. Oh, what an inestimable blessing to a boy to have such a mother and to value the letters so that here almost a century later they are found carefully kept; I suppose all she wrote, for postage was so high then that letters were not an everyday or a weekly matter.

August 24.

Reading with great pleasure the "Life and Letters of Washington Allston," and came upon so many bits of wisdom which I would like to keep, for instance : —

"Confidence is the soul of genius. . . . A little seasonable vanity is the best friend we can have."

"It was a saying of Alcibiades, and I believe a very just one, that 'When souls of a certain order did not perform all they wished it was because they had not courage to attempt all they could.'"

All this written by my great-uncle Washington Allston, August 24, 1801, to the artist Charles Frazer — to-day 109 years ago. We have a very beautiful miniature of him and it has the face of a wise man and almost a saint.

PEACEVILLE, *September 1.*

A beautiful morning, though clouds are still flying. Everything is fresh and sparkling after the rain.

Had a terrible temptation — a letter yesterday from A. C. begging me to join her in Columbia to-morrow and make the journey with her to Highlands, where my sisters and their families are. All summer J. has been urging me to spend the hot months with her, and sent the check for my travelling expenses, which I returned, as it seemed to me my duty to stay here.

Now the thought of that wonderful exhilarating climate

and beautiful scenery with all my dear ones was too much for me. I determined to go. Drove to Cherokee and gave Bonaparte directions for the conduct of everything during my absence, specially the curing of my precious pea-vine hay. Sent word to the ferryman to have the flat on this side of the river at 3:30 A.M. Had the shafts taken off and pole on the buckboard, so as to drive the pair, as I wished to take my steamer trunk with me.

It would be necessary to leave here at 2:10 A.M. to take the 6 A.M. train, though it is only eighteen miles; but the ferry represents an unknown quantity. After all was done I felt very light-hearted. To turn my back on heat and worry, on discouragement and continual effort, was delightful and I walked down to the barn-yard with light and springy tread.

On my way the gorgeous sunset struck me and I stopped, spellbound by its infinite beauty. Oh, the tenderness of the light, fleecy pink clouds; oh, the passionate red of the darker ones; oh, the golden glow of the horizon! Could anything hold more intense beauty and delight? Could one look at that and shrink from toil and moil and sweat in the path of duty? Was I a coward? Was I a shirk? Had I not chosen my own path and was I too much of a weakling to walk in it? Was I willing to leave the burden and heat of the day for two old darkies to struggle through alone?

I stood there filling my soul with beauty and strength until the last beam faded, then went to old Bonaparte and countermanded all my orders. It was all easily done except to notify the ferryman that he need not be ready for me, and I will send him a little present to make up for that. I have but one distress. I wrote to my sister by to-day's mail to say if it was possible to do it I would go, and I hate to think of her disappointment.

I drove back to the pineland in my little old rattling buckboard, it being too late to have the pole taken off of the other one, and a great peace filled me. Chloe was overjoyed at my change of plan, though she had encouraged my going in every way and had my trunk all packed. As for Goliah, he fairly glittered with joy, which condition was contributed to by the habit he has taken up recently of greasing his broad little black face very thoroughly with the vaseline I provide for keeping the harness soft. He seems to find infinite comfort from rubbing a quantity on his hands and then massaging his face very hard with both hands. It always amuses me, it is such an odd thing for a child to do.

September 4.

A glorious autumn day, really cold. I was very busy all morning bottling some blackberry wine made in 1903. Somehow there was great haste all day. At the plantation got very unhappy over the fear of cockspurs in the hay. It is impossible to make the negroes understand the importance of destroying them. Last year the horses had none of the best hay. It was all kept for the cows because there were a few cockspurs in it.

The scuppernong grapes are ripening very fast and are delicious.

September 6.

I certainly have been rewarded for not going to the mountains, for the mountain coolness has come to me — the weather has been perfect since the day after my decision. This morning Nat came from Casa Bianca looking more cheerful than usual. He told me he had nearly sold my cow Onyx.

I received this rather coldly with the commonplace as to a miss being equal to a mile in such cases. He said Mr. E. had come up with his brother to look at the cow and asked him the price and he answered $27.50, and that the brother took out the money and counted it out into his hands when Mr. E. came up and said: —

"I can make a better bargain with Mrs. Pennington than that. I'll write to her. Take back your money."

"Nat said: 'De money luk very sweet een my han's, but I gie um back to de gen'leman. Yu get letter frum em yet?'"

"No," I said, "but I can't imagine what made you say $27.50 for that cow with her splendid calf. I never said less than $30. I am glad that trade is off."

This was true and yet not true. I never would have sold Onyx for less than $30 myself, but if Nat had brought the smaller sum at this moment I would not have reproved him, as the constant call of the laborers to be paid presses on me daily.

After much wandering talk Nat took from his pocket a roll of money and counted it out to me, saying: "Ef I ain't succeed to sell de cow, I dun sell John Smit fu t'irty dolla'!" and there it was.

I was too thankful for words and yet sorry to part with John Smith, a handsome, long-legged Brahmin steer, who travels like a horse. However, as he is not yet three years old it is a fine sale and I praised Nat accordingly and gave him a dollar. Every time I think I am going to have a fine young pair of oxen I find myself obliged to part with them and be content with my faithful old ones, for there is always good sale for the steers even when nothing else sells.

At Cherokee I saw no sign of Gibbie, but was pleased to see the three milch cows tethered in the lush grass of the cornfield. I have long tried to get Gibbie to do this, in vain; it is not the habit in this country, and Gibbie is sure would have fatal results. What made him come to it I do not know, but I was delighted to see them knee deep in grass, evidently satisfied, for they were not eating, only chewing their cud meditatively.

I passed on to the house and after a while went down to the garden to see about the turnips. Just as I was about to cross the little foot bridge which leads from the barn-yard I saw basking on the plank a terrible looking moccasin. I turned away to get a stick long enough and strong enough to give me courage to attack him. When I came back armed the snake had disappeared, and I was about to cross when some instinct warned me to look well, and there just under the bridge in the flowers and grass that grew so luxuriantly as almost to touch it, I saw the beady eyes in the erect asp-like head fixed on me.

I summoned all my nerve and after a severe struggle killed the deadly thing. Even after I threw it some distance away with my strong staff it was hard to make myself cross the narrow bridge. I finally did get into the garden and found the turnips in great need of work.

On my way back I looked into the field where the cows were, and there was Moselle, my thoroughbred Guernsey, whom I had seen all right a half hour before, prostrate on the ground, her head under her! I flew through the gate and to her, to find that the horns were fastened in the ground, her forelegs bent over her head. The rope round her hind leg had evidently caught when she went to lie down or get up, I don't know which.

I called aloud for Gibbie, Bonaparte, Goliah, but no one came. Moselle's breathing was like a very loud snore. I tugged at her forefoot to lift it off from her head, putting all my strength, but in vain; when exhausted by the great effort, I called again and again, then getting no answer returned to the tugging.

At last I succeeded by a mighty effort, then another mighty effort, and I got her horns out of the earth and put her head in a natural position, when she lay as if dead. The terrible sound in her breathing had ceased, but she plainly said she was dying. I loosed the rope from her foot and from her head and encouraged her to get up, but she lay with closed eyes, and I left her, for I know how easily cows give up and die.

I went up to the house, where Goliah was putting Ruth in the buckboard, for it was sunset. Then Gibbie appeared and I told him Moselle was dying and reproached him for being away from his post of duty. When I drove out after all this expenditure of feeling Moselle was quietly eating as though nothing unusual had befallen her. I felt like Mother Hubbard after her trip to the undertaker's. Altogether it was a trying afternoon and I am very tired.

September 7.

Had some important writing to do this morning, but before I could begin Patty came in to say two "ladies" wanted to see me. I went out to find Totem's two daughters, who wished to get me to protect their property for them.

They said their father's "stepwife" had advertised the horse and buggy and cow and calf for sale, all of which things had belonged to their own mother and the "stepwife" had no right to sell them. I spent the whole morning talking to them and writing for them to the Probate Judge and others.

Patty came in. Alice R. Huger Smith

Totem was a faithful servant and their mother an excellent woman, and I shall do all in my power to have their property protected. At the same time I tried to make them understand that the "stepwife," having been legally married to their father, however short a time before his death, had a right to a proportion of his property.

As soon as they were gone I went to the plantation, where terrible havoc has been made in the corn by three hogs belonging to negroes who live miles away in the woods. It is a most difficult thing to get any redress for this. Bonaparte asked me to walk through the corn-field to estimate the damage, and really I think one-third of the corn has gone.

I cannot believe it is altogether the work of animals. I think they have been assisted by humans, for while great quantities of corn stalks are bent down and you can see where the corn has been eaten on the ground, in many, many cases the stalks are standing straight up and the ears are gone. However, I say nothing about that, as it would be useless.

One of the hogs I hired a man with a dog to catch two weeks ago. It weighs over 200 pounds and the man charged $2 for catching it. I have fed it in the pen for seventeen days. Now the owner, a very well-to-do darky who has a pension from the Government and is above work, says he cannot possibly pay $7, which is the amount I fixed upon, though the damage is much greater than that, indeed, four times that.

There are still two smaller hogs of about 100 pounds each in the field.

I have a strong wire fence around and I cannot help thinking the hogs have been let in at the gate. Of course a man would have them shot, but I cannot do that.

The milk is falling off, and to keep up my butter engagements I will have to stop sending the pint of milk daily to Eva which I have been sending for six weeks. She is Gibbie's mother, and when the doctor said she should have fresh milk I gladly gave it, but she is up and about now, and if Gibbie will not take the trouble to take the milk from the cows because he is in such haste to go out hunting, his mother will have to suffer along with me.

Bonaparte and Kilpatrick are working on the flat which needed overhauling and repair. It is a heavy expense, but as the rice is doing well there must be a dry, tight flat to bring it in.

September 10.

Very hot again. This morning I was working at a serenade of Rachmaninoff's when I fainted. Good Chloe got me on the lounge and dosed me with ammonia and I got over it, but could neither write nor read without a return of the terrible feeling. So I had the room darkened and kept quiet until 4:30 when I had to go to the plantation.

September 13.

Miss Penelope sent me word she was unequal to going to church to-day. So I had to play the organ as well as sing. Though a little rickety still, I enjoyed the organ in its rejuvenated condition. It is very sweet and full. A beautiful sermon. Thanked the good Father for his many mercies. The Sunday-school children came promptly at five and were most interesting.

September 14.

Making wine from the scuppernong grapes; ten quarts of grapes made two and a half gallons of wine. It is a very simple process, and yet the wine is very nice. It would make most delicious champagne if we had strong enough bottles to put it in at the right stage, but it bursts ordinary bottles, so we leave it uncorked until that stage passes.

I make it because I find a portion of wine is a most acceptable present to the men of the family at Christmas time — only it must not be too sweet. The scuppernong grape grows so rapidly and vigorously in this soil and climate that it would be worth while to plant it largely for transportation to places where wine is made. In this State it is under the ban, but there is no law to prevent sending out the grapes.

Every negro cottage through the long line of villages which fill the pine woods has at least one scuppernong vine, from which they sell bushels of grapes, besides eating them for a month. One vine will cover several hundred feet of space, for they are never trimmed, but grow laterally on scaffoldings made about five feet and a half from the ground.

They do not grow in bunches like other grapes, but only four or five very large grapes together, so that when you go under an arbor of ripe grapes you see no leaves above you, only a canopy of grapes, the leaves being all on top, and there is no more delicious experience than a half hour under a really old grape-vine in early September.

The older the vine the more luscious the grapes, and the perfume is most exquisite. It is a native of North Carolina, but takes kindly to this State and requires no spraying or care of any kind beyond breaking away the dead twigs and branches during the winter season — and mulching with dead leaves.

September 15.

Had a present of a bushel of grapes from Old Tom's children — such a pleasant surprise! The grapes from my arbor are so enjoyed by the whole plantation that I never get more than a peck at a time, so that it is a great thing to have such a handsome present. Presented the bringer with a dress for herself and shirt and cravat for the brother. That is what a present means with us — good will expressed, and a handsome return.

PEACEVILLE, *September 16.*

This morning had a delightful present of venison. S. M. killed a deer yesterday.

Sent Chloe, Patty, and Goliah to plantation to pick the

last of the grapes, and I tried to refresh myself by reading "Peter." Yesterday when I drove down Ruth behaved in the most unaccountable way. I had S. R. with me and we were driving up the avenue to the barn-yard, which is called the Red Bank — I do not know why, as it is not a bank nor is it red, just an avenue bordered by live oak trees with the fields of corn, peas, potatoes, etc., beyond. The growth is very luxuriant and thick on each side under the trees.

About halfway from the gate Ruth suddenly shied violently, shivered and shook, and though the road is quite too narrow to turn she backed violently right into the ditch, and before I understood what she was doing she had turned the buckboard around most cleverly and was rapidly on her way back to the gate with every sign of terror. As soon as I realized what had happened I drove into the field on one side of the road, turned and drove back up the avenue toward the barn-yard, the road she has travelled all summer every day but Sunday without showing the least fear of anything.

I made Goliah walk ahead until we got near the spot which had so terrified her. When I saw the fit of terror returning I gave the reins to S. who fortunately was with me and is a very good whip, and I got out and led Ruth by with the greatest difficulty. I do not know what to make of it unless there was some one hidden in the ditch who was very obnoxious to her.

The only time I ever knew her to shy so violently before was once when I was driving down to Casa Bianca alone. In a perfectly open, clear road, with a deep ditch on each side, no bushes or underbrush at all, she was trotting along briskly when suddenly she made a terrific shy to the right and bolted. In a few yards I pulled her down, and wondering greatly at her conduct I looked back to see if there were any stumps which I had not noticed, and out of the ditch on the left side of the road rose a most fearful looking head, a white man's, all overgrown with hair, hatless, dishevelled — no doubt a fugitive from justice who had wandered the roads a long time, from his aspect.

Needless to say I did not tarry to ask questions, but let Ruth travel at her very best speed, and that evening returning home I drove as fast as I could, whip in hand, but had no further trouble with Ruth.

On this occasion surely if there had been any one hidden in the ditch Don, the setter, would have found him.

Coming home she still seemed nervous. Goliah says "plat eye" and Chloe says "speret, Miss Pashuns. You know Cherokee is a ha'nt place, dat Red Bank road speshul, en wen yu cum to de Praise House lane dat dem home.

"T'ree time dem 'tack me dere. One time I bin a cum f'um Nannie weddin'. I see a man walk right befo' me, en I call to um en say 'Elihu! Dat be yu? Wait f'r me,' en de man neber answer, en w'en 'e git to de gate 'e neber open um, 'e jes' pass trou' wi'dout open, en den 'e tu'n 'eself unto a bull, en rare up befo' me. Den I kno' 'twas plat eye, en I say to meself 'Trow down yu fader h'art, en tek up yu murrer h'art,' 'en I dun so. 'Kase yu kno', Miss Pashuns, yu' murrer h'art is always stronger dan yu fader h'art.

"Oh, yu didn't kno' dat? 'Oman h'art is stronger dan man h'art w'en yu cum to speret en plat eye. Yes, Rut' see dat same man en I jes' t'enk de Lawd she ain' cripple yu."

That night she returned to the subject and told many wonderful ghost stories, all of that same road, and said Gibbie was so afraid of going along there in the dusk and reminded me that he never would wait to take my horse when I was out late, and that was the reason. As I still pooh-poohed her stories she put on quite a superior air and said : —

"Critter kin see mo' dan me, Miss Pashuns, en I kin see

"Plat eye!"

mo' dan yu, fer all yu kno' so mutch mo' en me. W'at I tell yu, 'tain't wha' I hear, but wha' I see meself."

There is no doubt something in what Chloe says about creatures, as she calls animals, seeing more than human beings. There is a spot on the road about a mile from Casa Bianca where a man was killed by a fall from his horse, which shied violently, throwing him against a tree. This was about sixty-five years ago, and though it is now a commonplace looking spot enough, my horses rarely pass it without shying.

September 18.

Yesterday I gave Gibbie a severe talk because of his total neglect of his work — the stables not cleaned, no pine straw hauled for bedding, the calves starved, yet the cows only half milked. I would not mind losing the milk so much if only the calves got it, but they look miserable, especially Heart, the little Guernsey I so wish to raise.

He is intoxicated with the rice bird and coot fever and spends every night out hunting, and of course in the day he is too sleepy to do anything. He answered almost insolently for the first time, for usually he has the grace of civility.

September 22.

Went down after early dinner in great haste to peas field prepared to help pick out cockspurs, but found that Gertie and two other women had finished. I went over it prepared to find it only one-half done, as usual, but to my delight found it thoroughly done. They had two large barrels packed tightly with cockspurs, root and all, the burrs being still soft; and look over the field as carefully as I could I found not a single plant. I had the pleasure of praising them warmly. It was Gertie's doing, I know, as Bonaparte put her in charge.

Chloe returned to the subject of "sperets" to-night and

would insist on going back to all the strange things that have happened in her experience. There is no doubt that Chloe would develop into a most successful medium if she was in the way of knowing anything about the present craze for spirit manifestation.

She called to my remembrance one very strange circumstance that took place the first year I was alone at Cherokee after my dear mother's death. It had been the habit of our household to have family prayers, and when I was left alone I determined to continue the evening prayers and for that purpose had Chloe come in at 10 o'clock.

It was curious how reluctant she was to have me act as home chaplain. She evidently did not consider me equal to the situation. However, I made a point of it and she graciously came.

After prayers were ended she would stand at the door looking very dignified in her white head handkerchief and white apron and talk over the events of the day, the condition of the poultry yard and the evil deeds of the generality of mankind. This little chance to tell her trials and tribulations was greatly enjoyed by her, and I tried not to be impatient at the wealth of detail, and impossibility of getting back to my book, for I knew that alone made her consent to come in to the little service which meant so much to me as a survival of the past.

One particular evening when she was in full swing I was sitting in one arm-chair by the fire, the other being empty, and on the rug stretched off in front of the fire asleep lay a very handsome Skye terrier which had been recently given to me as a protection, my dear little old black and tan Zero having died that summer. Suddenly Blue Boy woke, rose, every hair on end. He growled, he sniffed, he snorted, and then made a dash at the empty chair, barking furiously.

I tried to pacify him, called him to me, patted and petted

him, all in vain. He got under my chair, but he continued to bark fiercely. Finally I was annoyed by it and got up and sat in the empty chair. It meant nothing to me but that Blue Boy had had a bad dream.

I went on talking to Chloe and as Blue Boy quieted down and went back to sleep on the rug I got up and in my impatience at the prolonged talk began to walk about the room, I was so anxious to get back to my interesting book. In a second I heard a growl and Blue Boy was on the rampage again, more furiously than the first time. He attacked the empty chair, making a dash to within a foot of it and then running away, only to renew the attack.

I was quite provoked and was going to slap him when I looked at Chloe. She was white almost, with a look of terror.

"Miss Pashuns, 'tis Ole Miss' Blue Boy see."

"What nonsense, Chloe! You know that is impossible, and even if it were possible, why should Blue Boy bark at mamma? You know all the dogs were devoted to her."

Chloe answered: "Miss Pashuns, you fergit, you git Blue Boy since Ole Miss' gone; him 'oodn't kno' Ole Miss'."

It ended by my taking the dog up and carrying him out of the house. Up to this time he had always slept in my room at night as Zero used to do, but when I was ready to go upstairs that evening and called him he would not come inside the door. He wagged his tail quietly and licked my hands but refused to come in, and from that time I never could induce him to stay in my room either night or day. He would lie on the rug until I was ready to go upstairs, but then he went to the front door and insisted on remaining on the piazza for the night.

After putting Blue Boy out I returned to try to reassure Chloe, who was greatly agitated. I told her that if the Good Father, in whose hands I felt so safe, should see fit to let those whom I so dearly loved in the flesh, return in the spirit to watch over me in my lonely life, it would make me very happy, and that I could not understand it being a cause of terror to any one.

"But," I said, "I do not feel called upon to decide whether that is possible as our world is constituted. I only have a firm abiding faith in the mercy and love of God and in His determination and ability to keep all those who put their trust in Him and walk in His commandments."

Then I went to the piano and had her sing with me that beautiful old hymn, "How firm a foundation, ye saints of the Lord."

CHAPTER VII

PEACEVILLE, September 18.

WENT out to the mission in the pine woods with Mr. G. Quite a good congregation. They all walk miles, and bring their babies. Saw a most forlorn specimen of a man, sallow, emaciated, miserably clad, with three children wrapped in a heterogeneous collection of garments. Mr. G. turned to me and said: —

"You know Mr. Lewis, Mrs. Pennington?"

Before I could answer the poor gaberlunzie spoke up and said: "Oh, yes; she stood for these," waving his hand over the thin little objects. "You 'member, Miss Patience, this is Mary Frances and this is Easter Anne and this is Thomas Nelson."

I never felt more abashed in my life. Such a party to be responsible for! I stand for so many, many poor little babies, for whenever there is a christening I am in demand; but I never have had such a forlorn little company as this on my soul.

As soon as I recovered I asked how it was that I had not seen them for so long.

"We've bin a-travellin'! We moved off for a good many years now, an' that's why you've sort of lost us."

I asked where his wife was.

"Well, ma'am, she's gone; got tired of the job, an' lef' me."

"And who cooks?"

"Mary Frances cooks."

"And who washes the children's clothes?"

"Mary Frances, she washes, an' Mary Frances, she mends an' does everything."

When I looked at the wizen little girl, with her sallow blue skin and her skinny little arms and hands, I could scarcely keep back the tears, but I spoke very cheerily to her and complimented her on the get-up of the family, which truly showed ingenuity.

She told me she was 10 and Easter Anne 8. I could scarcely believe the tiny child was 10, but I promised to make some clothes for them before the next day the rector came, which will be Sunday three weeks. She did not seem excited or even pleased, but answered "Yessum" in a listless voice to all I said.

I asked some of the other people about them and found there was great indignation about the wife. These people are severe on the erring; it seems necessary to their self-respect.

September 19.

Bonaparte has been away on a little vacation and I have been superintending all the work personally for the past two weeks, and it is impossible to get a decent day's work done. The women just scratch the ground a little with their hoes when your eyes are on them, and as soon as you allow yourself to be diverted for a moment they stand quite idle.

September 21.

Was telling Miss Pandora about the Lewis children and how I was searching all my possessions to find something I could cut up to make into clothes for them. She said at once: —

"I have the stuff I got some time ago for a skirt. I will send it to you to-morrow."

I remonstrated, telling her it would not be suitable, as they should have stout stuff for clothes; but she persisted and sent it, three and one-half yards of very pretty crash. I nearly sent it back because it is too thin and unsuitable and would make such a pretty suit. However, after much consideration, I determined to offer it for sale, and if I succeed in getting the money for it, I will spend it in homespun and calico to make up. This afternoon I took it down the village and showed it to several people, and I finally left it to be examined.

September 22.

My little trading effort has been most successful. This morning I had a note to say that the stuff had been bought and sending me the money. I at once went down to Miss Penelope's and bought fifteen yards of stuff, different kinds for the different ones; and then set to work to cut and make three little frocks. Patterns seemed a difficulty, but I would allow nothing to cool my ardor. I made my own patterns, for these pine woods people know nothing of fashion in children's garments, and I am making them as I used to make children's clothes long ago.

The draperies Mary Frances had hanging around her were down to the ground and so were Easter Anne's. It will no doubt be a shock to have these only reach their ankles, but they will have time to get accustomed to it before cold weather comes. One wonders stupidly over things out of one's own beat, as it were, but of course when children do not have shoes and stockings in the cold weather trailing garments are preferred.

My neighbor the widow asked me to let her do some of the machine stitching for me, which is very nice of her, my machine being out of order for the first time in its thirty-seven years of service. I think Patty must have been experimenting with it, for it did beautiful work the last time I used it. Let no one turn up her nose at this old friend and say, "At least in machinery new friends are best." We are a faithful hearted people down here and see the beauty in our old friends, even though aware of the pathos of increased effort.

September 23.

This morning went over by invitation to look at the widow's steers which she is to sell to-day for a good price. They are very fine and perfectly gentle. She is a wonderful woman, doing all her own work and so much of it. Her vegetable garden has not a blade of grass. It contains turnips, cabbages, carrots, beets, and tomatoes. She milks her cow herself, waters her great number of flowers, drawing the water from a well with an old-time arrangement; keeps her large rose garden in order and has the house filled with fresh flowers.

To-night I finished two of the little frocks, and they look very sweet. I could not help stitching on a little band of contrasting color. Children's clothes should be pretty; all things connected with childhood should be pretty. The little ones thrive on things that feed the eye with beauty. The Great Father teaches us that wherever we turn in the loveliness spread around us everywhere, but we are so slow to learn.

September 24.

I undertook to have Jim do some mowing in the neighborhood, there being difficulty in getting a mowing machine for hire. But yesterday when the field was about half cut the blade broke, and now I have to send off eighteen miles to get another and by the time I have done I will be on the wrong side as far as profit goes.

Sunday, September 25.

Goliah came to me in great distress, weeping and saying his hoop, which he hung on the fence, was gone. I told him some of his very rude companions, whom he occasionally brings into the yard, had taken it.

"No, no," he said. Some one had broken it up, and he thought it was Patty. I reproved him for supposing Patty would do such a thing, but later when he had gone out of the yard I asked Patty if she had troubled the hoop. She said, "No." I answered, "I am very glad to hear it, for I would have been very angry if you had destroyed Goliah's hoop; it is an innocent amusement and keeps him out of mischief."

She went out quietly, but I soon found the yard was in the greatest excitement. Goliah returned and found some other cherished possession gone, and he sat on the back step and cried and sobbed. I tried to quiet him, but in vain, and then to add to the tragic effect his nose began to bleed and his clean white shirt had great splotches of blood.

Goliah cried and sobbed.

There raged a tempest in a tea-pot this blessed day of rest. I could not stand it, and ordered Jim to put Ruth in the buckboard and gave the whole yard a holiday. I told Chloe I would not have any dinner, so she could go to visit her family. I was going out to St. Peter's-in-the-woods to take the clothes I had made to the children.

So I escaped and went to church, and then had a lovely drive through the pine woods, and the joy of putting the frocks on the children and finding that they fitted nicely, only I saw they thought them too short, so I said I would take them home and make them longer.

The wandering mother had returned and I had not the heart to look harshly at her; the poor little ones looked so happy — not a change of garment or any other change, just the little gray faces, which had looked so lustreless and lifeless, were full of interest and animation. The poverty of the surroundings, the doorless hut with no attempt at furniture — it was all pitiful.

It is very rare to see such poverty in this part of the world. I have never seen such a case before; but the man is a semi-invalid and work in the field for the woman not easy to get, I suppose.

I did not remember what a long way we have to go when going home. I had not started early, for I went to church first, and then went to ask a friend to go with me. At any rate when we had gone about half of the nine miles home the swift, soft darkness fell. It was a perfect evening and we were enjoying the delicious cool of the night air when I looked ahead in the very narrow road, a deep ditch on each side, and saw a steady bright light coming. I knew it was the one danger I feared.

Just then my companion saw it. "Patience," she said, "that is an automobile; the doctor's, I know. There is an ill man out on this road; what shall we do? He cannot see us."

That was perfectly true; we were completely in the darkness, and his big light did not cast far enough to give him time to stop his car when he saw us, and the road was too narrow for two buggies to pass, without great skill in driving.

I drove steadily on, but I felt dismayed. There was, I thought, not far away a bridge of pine saplings across the ditch on the right. If we could reach that before we were too near we might escape.

Meantime my companion said, "Let us call aloud, they may hear." So she lifted up a splendid strong voice and called, and when she ceased, her voice exhausted, I took it up; but on, on came that star of fate; it had the most curious inevitable look.

Only by its growing larger and larger could we know it

was moving. Finally when A. said: "You must stop, you cannot go on," I knew she was right and that I must stop without having reached the little bridge which meant safety.

I stopped. On, on came the glare. Ruth, like myself, seemed fascinated by it. We were so powerless, for now we could hear the roaring and knew our voices were impotent to reach the driver. There was not fifty yards between us, and on they came. No, there is a change in the sound. They have stopped! Thank God! Thank God! It would have been a grizzly, grinding death.

The driver leaped out and came to us, white as a sheet.

"Oh," he said, "just in time! Miss A. saw your white shirt-waist and said, 'Stop: there is something ahead!'"

He was just as good as gold, and when I said if he moved the auto to one side a little I would undertake to lead Ruth by, "No, no," he said, "we must find some other way, the road is too narrow."

I told him of the little bridge and he found it just between us and the auto, and he insisted on leading Ruth over it; turned out his lights and glided past quietly, and then led the wonderfully well-behaved Ruth back into the road, and with hearty handclasps and thanks we proceeded on our way. Very thankful hearts beat within us and the mercy and goodness of the Great Creator seemed to be shouted to us from each brilliant star above.

September 26.

Started the flat off to Gregory for the fertilizer, five tons half lime and half 8-1-4. The river winds so that by water it is twenty-three miles about, while by land it is only fourteen. If everything goes well, they should get back here by Thursday. R. came down to spend three weeks with me, and he is helping me prepare the land for the alfalfa. It is so delightful to have him. He finds the nigs very trying. Yesterday he spent a good deal of time fixing a harrow, which was too light, by wiring on to it securely a long iron bar to make it heavy enough to crush the sods. Finally he got it just the right weight and started Elihu to work with it, and was delighted with the results. To-day when he went down to the plantation he found Elihu harrowing, but without the bar; he had cut every wire which had been securely fastened and taken off the bar. When R. asked him why he had done it, he scratched his head and laughed and answered : —

"Jes' so, Mass' Bob."

September 27.

The corn is all gathered and has done very well — 814 bushels of slip-shuck corn on seven acres. Gibbie is very proud; he feels that he and Paul, the single ox, have done it all.

September 28.

Went to Casa Bianca and walked around the banks. The little rice planted looks fairly well. Nat seems to be doing his best in face of much opposition and difficulty. On the way back stopped at Cherokee and found that Elihu got back with the flat of fertilizer at sunrise this morning, which was doing splendidly. It was most fortunate, for this afternoon the storm signals are out in Gregory. Mr. L. was afraid to leave town with his tug towing a lighter, so it would have been impossible to bring an open flat out.

October 2.

The fertilizer has been distributed over the alfalfa field and the whole field is in fine order. Now the delay is in the nonarrival of the seed. I have sent to the railroad station several times, but they answer firmly that it has not come. It is very provoking, for all the books say it should be planted not later than October 1.

October 6.

R. was obliged to leave to-day, and without helping me plant the alfalfa, as it has not yet come. It is too bad, for it would have been such a comfort to get it in while he was here. I asked him to go to the station very early to-morrow — the train leaves at 6 A.M. — and ask permission, very politely, to look through the warehouse himself for it; he seemed to think this an unusual and unreasonable request, but I know the ways of the freight office in Gregory so well that I am sure the alfalfa is there.

October 7.

Elihu returned at one o'clock, bringing the sack of 100 pounds of alfalfa and a note from R. He had asked for it and was told it was not there; then, politely, he asked if he might look through the warehouse; permission was granted, and almost the first thing he stumbled upon was the bag. When he told the man in charge he had found it and pointed it out, he looked at it and said: "Oh, that's it, is it? That's been here two weeks."

I called Bonaparte at once and used what was left of the culture R. had mixed, though I felt uncertain as to whether it was still good. It was only enough to moisten a half bushel which I had well stirred and then spread out on the piazza to dry. Then I proceeded to put together the stuff for a second lot of the inoculating liquid. I had had the packages quite a while, and felt anxious about it. It was in proportion for ten gallons, but I only mixed five, putting in half of the package of each instead of the whole. That is the worst of being so remote from everything — the difficulty of replacing things if anything goes wrong. Whether the tub leaked or the culture evaporated, I do not know, but the quantity R. mixed should have been enough for the 100 pounds, but it has vanished or rather "minished," to use a very pregnant negro word, and now I have to use these old ingredients.

October 8.

Put in the second ingredient of the culture, then got Bonaparte and two boys with the seed drill, which I had asked R. to rent from a neighbor, and proceeded to plant the alfalfa seed wet with culture yesterday, as it was quite dry. The Imp and Manuel were charmed to run the little drill and fought over it, for I would not let either one do more than six rows.

A glorious sunshine, thank the Good Father. I hope I will get the cotton picked to-day and a good many peas, too.

10 P.M. A fine day's work. Took Patty and Goliah in to pick peas, and they did well and enjoyed it. I hear on one has made any peas this year, but I have made a great quantity, which is a great mercy. Patty, Goliah, and I picked peas along with the other hands.

Lizette was there with her little baby, the first time she has had it in the field. It is tiny and sits up very straight and looks like a little black doll. Her little son Isaiah sits and holds the baby all day. I constantly intervened and had its little head kept from rolling off, as it seemed likely to me to do when it was asleep.

I told Lizette about the children in the East Side Settlement House, each baby so comfortable in its basket, with no danger to its little delicate spine. Then as that did not seem to attract her I told her of the Indian babies safely bound to a straight board and hung in the trees. That desperate cruelty, as it seemed to her, roused her to speech, which it is difficult to do. With great indignation she told me there was no need for her to be so cruel to her baby as she had a boy to mind it. The boy may be four, but I do not think he is quite that. I am going to make a nice little box, with a handle and a little pad in it for a mattress, to carry the baby in.

I enjoyed every moment of this beautiful day drinking in

God's glorious handiwork of air and sky — everywhere masses of goldenrod and banks of feathery white fennel.

October 10.

This morning Miss Pandora brought me a present of a dozen splendid apples! I was greatly touched by it — such a great present here, where we see no fruit but pears. It was Miss Melpomene's birthday and I was busy fixing up a little offering for her, a summer duck nicely roasted (for Chloe's cooking a duck doubles its value) surrounded by tomatoes from my pot plant, which are supposed to be very superior in flavor. I sent a note asking Miss Melpomene to go with me to Cherokee this afternoon prepared to pick peas.

She seemed startled but accepted with pleasure, and when I explained that she was to keep all she picked she was charmed, as hers have failed entirely. I drove to the field and left her there, having lent her my pea picking apron. It is made of light blue denim, quite long and turned up like a sewing apron only much larger, for it can hold nearly a bushel of peas.

I drove to the barn-yard to leave the horse and buckboard and return to help her pick, but I found ten hands waiting with huge bundles of peas. Bonaparte said with great impatience, "Dem do' want no money, dem want peas," so I said at once, "I don't blame them, let them have the peas."

But I had to stop and make the necessary calculations for each to get one-third of what she had picked. It was quite a business, for in all they had picked 1197 pounds of peas, some picking 150 pounds, others only fifty. They are selling for 10 cents a quart now, so naturally the pickers prefer taking a portion of the peas to money.

It was nearly dark when I got through and went back to Miss Melpomene, who thought something must have happened and seemed to think she had picked quite too many peas and was eager to make me take some. It was an original birthday party, but we both enjoyed it greatly, and the drive home was delightful, and we were very gay.

October 11.

Had Eva to sow by hand the little of the inoculated seed left yesterday. Assisted by Bonaparte I mixed the rest of the seed — one and three-quarter bushels — with the liquid culture and then spread the wet seed out in the piazza to dry. The stuff smelled very yeasty and queer. I do hope it is all right. As I had much more liquid than I needed, I mixed it with earth so that I may use it in future.

Yesterday, with a storm coming up, I was unable to get any one to haul in my beautiful peavine hay. A month ago Gibbie had asked permission to be absent to-day and I promised him he should go. I sent word to Elihu and George to come and handle the hay, but there was a funeral, and not a single man could I get.

Had Eva to sow by hand a little of the inoculated seed.

October 12.

Drove down early to Cherokee, and finding the seed dry drove rapidly to Mr. L.'s place to get the drill, but instead of using it yesterday they were sowing rape to-day, so there was nothing to do but return quickly, send for all the women I could get, and sow it out by hand. The sowing was easy enough, though slow, for the women are accustomed to sow rice by hand, but the covering was the difficulty. I had eight hands all the time and then when the hands who were picking peas knocked off I called them in to help. The moon was high in the eastern sky when the last row was sowed, and then we had to stop, though about one and a half acres were not covered. It had been a great rush, and the hands all worked well and I paid them extra, for though they had not started till late, as I had counted on getting the drill, they had worked steadily. I was completely exhausted when I got home.

October 13.

When I got to the plantation this morning I found Bonaparte had five hands covering the one and a half acres left uncovered last night, and they took the whole day, and it was abominably done. He was in a very bad humor and would not follow my directions, and give each hand ten rows for him individually to do, so that one could see who was doing good work and who not; but insisted on laying off with stakes a section for each one, saying the rows were too long, and he must keep them together and watch them. "Dem's too striffling for tek dem long row. I 'bleege to keep dem close togeder, so I kin watch dem. Dem's striffling no 'count, good f'r nutting," etc., ad lib. I simply had to leave the field or have a tremendous flare up, so while I could control myself I left; but it was very trying, for this is the richest part of the field, and he had got the hands in such a bad humor that they were positively digging the seed out of the ground instead of covering it. For a few seconds I was on the point of ordering him out of the field, but that meant destroying his prestige and authority for all time, and he has all the barn keys, and I believe is faithful to the trust; he is just mulishly cantankerous sometimes.

I found Gibbie diligently running the mowing machine cutting down the second pea-field, while the hay which was cut down Monday and Tuesday and had two solid nights rain on the piles was dry on top and steaming wet underneath. I stopped the mowing and led Gibbie from cock to cock and made him toss and turn the pea-vine hay while I sent George to do the same to the broom-grass hay. No one seems to have any sense. I told them to keep turning it as it dried and then to begin hauling into the barn and to try to finish getting it in to-morrow. I shall not be able to come down to-morrow, as I have to send for and entertain our rector.

Sunday, October 15.

The blessed day of rest is most welcome. It being the third Sunday in the month, we had our rector. His sermon, an excellent one, on the text "For every idle word," etc., struck the little congregation with dismay. As I came out of church some one said to me: "Do you think Mr. C. has been hearing anything about us that made him preach that sermon?" "No, no," I answered, "I think not, but I feel that it was specially inspired for my benefit." "No, indeed, Mrs. Pennington," another put in, "not for you, but for me." And so there was a group of self-convicted sinners, whose sins of the tongue had been brought home to them. As the rector went into my sitting room he laid the fiery roll on the table, and when he left the room I took it up to get the chapter and verse of the text, so as to look it up, and on the cover I saw written "First preached, August, 1888."

That afternoon a lady came to see me with a solemn, pained aspect, and after the usual inevitable complimentary prelude cleared her throat and began. "It is with much sorrow, Mrs. Pennington, that I state from indisputable authority that during his last monthly visitation our revered rector heard from a lady, who shall be nameless, things concerning some of our most respected families which induced him to give us the extraordinarily clever and appropriate discourse to which we listened this morning." It was very hard for me to wait politely for the end of this well turned sentence, and as soon as I decently could I answered with delight: "I am very glad to be able to tell you, my dear Miss Arethusa, that you are entirely mistaken. That sermon was written and preached first in 1888! It only shows that human nature is much the same at all times and in all places, for you are not the only one who thought its application personal."

The hymn singing to-night was specially hearty and Mr. C. seemed greatly to enjoy listening, which is rare, the measure of enjoyment being generally in proportion to the vigor of one's individual efforts.

October 16.

Yesterday I had planned to go over with Mr. C. This morning he asked me to go and promised to drive me down to see Mrs. S., who is 86, and I have been suddenly seized with a great desire to visit her. I have never seen her since my father took me as a child to visit her. She has lived alone on her plantation for many years, as I do, and though it is only about twenty miles away the getting there, crossing two rivers and then a long drive, is intricate. Last night it seemed easy to cross the rivers with Mr. C., spend the night with Mrs. C. and himself at the All Saints' rectory and go on the next morning, returning here Wednesday evening, but this morning I am discouraged and cannot go. I found Mr. C. unprovided with the medicines we think necessary to have on hand in the country, as he is a new-comer, so I put up phials of quinine, calomel, and soda and it took some time.

Sunday, October 20.

No service in the little church to-day. Sent to ask A. if she would dine with me and drive out in the woods with me afterward. I called Chloe and Patty and Goliah in and read the morning prayer and the beautiful hymns for the twenty-first Sunday after Trinity. I played and had them sing the chants and we had a pleasant little service. I always like to have a scriptural quorum.

I hope the Good Father did not mind my sewing a little on Mary Frances' frock after I had read the prayers. I was careful to do it in private for fear of offending a weaker brother.

We started out in the buckboard at three, taking the three little frocks for the children and a nice dark calico shirt-waist suit for the poor mother. The drive was charming. Stopped to see Louise M., who is so faithful in trying to carry on the Sunday-school. Her little log cottage was as clean as possible and she showed with great pride their potatoes just dug; she and her husband insisted on giving us some; they were very large, some of them weighing two pounds.

Her little log cottage was as clean as possible.

Went on to the Lewis's; found them very cheerful and just eating their midday meal. I went into the hut and so saw what it was, a very large spider full of hominy. That seemed the only thing, but they were perfectly content, their hunger being appeased by the abundance and heat of the meal, for it was steaming, not cooled unnecessarily as our food is by being transferred from one receptacle to another. The spider had the place of honor in the middle of the table. Each one was helped to a pan of hominy from it, and then retired out of doors to eat it.

They were all delighted with their frocks. I had collected some few men's garments for the gaberlunzie who owns the flock. But when I produced the calico frock for the wife she just overflowed with joy like a child. After many expressions of delight and satisfaction she retired to a corner to put it on, saying: —

"I'm sure, Miss Patience, no one could say I'm not a-needin' it, fo' I ain't had a chanct to wash this frock I got on till there comes a red-hot day, fo' I didn't have a thing to put on w'ile I'm a-washin' it."

When she appeared in it she swelled with pride and said : —

"The pusson that made this frock must 'a' measured me w'en I was a-sleepin'. No dressmaker could 'a' fit me so well."

I found that this poor soul had been for a week nursing a neighbor night and day, only leaving her long enough to walk the mile home and get her meals.

Mrs. Sullivan is very old as well as very ill and very poor, so that all the lifting and cooking and work of every kind Mrs. Lewis has done. When I said, "But you ought to get your meals there," she answered : —

"There isn't enough, Miss Patience, in the house but jus' for her, an' I'm thankful that we've got plenty o' grits to eat now; nobody need be hungry here."

It certainly is a lesson in more ways than one to go among those whose lives are so elementary. This woman, who has been accused of failing in her highest duties, who knows the daily presence of want, who has never had enough of anything but air and sunshine and the breath of life, spends day and night and all her strength in nursing a woman for the moment poorer than herself, in that she is old and helpless, and there is no feeling that she is doing anything unusual.

She put some of the dry hominy in a bucket and saying, "Now I mus' be goin'; Miss Sullivan begged me pitiful not to stay long," she took the bucket and started off at a brisk walk, but I asked her to sit on the back of the buckboard as I had to pass the house. This delighted her and we had much talk.

I asked if Mrs. Sullivan had no children who could help with the nursing. She said she had two.

"Yes, mum, she has a daughter, but she's mighty feeble an' she lives three miles away, an' it jes seems as if she couldn't get to cum to her mar; an' when she does git there, well, she's that tuckered out an' that sorry fur her mar, that she jes sets in to cry. Then Miss Sullivan's son lives with her, but he seems as if his mind was a-goin'; he kyan't do nothing."

"Doesn't he work?" I asked.

"Oh, yes, mum, he goes out an' works turpentine — that's all they've got to live on — but he don't think to cut a stick of wood, or bring a drop of water 'less'n you tell him to do it. His mar's too sick to tell him, an' he'll jes sit there an' see the fire go out an' never think. But soon as I tell him to cut a piece o' wood he'll do it right off. He's a big strong man an' they say a powerful fellow to work, but he don't seem to have no head to think."

I was sorry it was too late for me to go in and see the old woman and her son and find out what was wrong with the latter. I remember John very well. When I taught him in Sunday-school he was a very mischievous boy, but not stupid at all. I cannot think what has come to him.

The drive home was delightful. No automobile disputed the road with us this time.

PEACEVILLE, October 27.

Had a message to-day from the man who rents the farm that he was ready to deliver his rent in kind. It was a little misty, but I had said I would be there, so rode out on horseback while Gibbie drove the ox wagon with the big rack to receive the corn and fodder. It is "much ado about nothing," but I was solemn and put down the little numbers in my book as Mr. C. measured, though I did not dismount. The corn, fodder, and peas could all be carried in the wagon at one time, so one knows it is not a fortune. We have about 1800 acres of wild land, and in the middle of the tract is a little cabin where my father's stock-minder used to live, and every summer as we moved from the plantation, all the cattle were driven out into the pineland to spend the summer, and fatten on the rich grass of the savannahs — perhaps that is why we never heard of Texas fever among the cattle in those days. Certainly the imported stock which my father always kept throve and flourished, and they returned in November fat and hearty. Now it is impossible to do this for fear of their being stolen, and the summer on the plantation is hard on them.

The ride was delightful, the mist so soft and caressing. This has been a perfect autumn season, and I feel like clinging to the skirts of each day of crisp, cool temperature and glowing, gorgeous color. I want to keep winter off as long as possible.

November 6.

As I sat on the piazza to-day about noon a runner came to the step, an unknown negro. He looked exhausted, and I said: "You seem tired; sit on the step."

"Yes, Miss," he said. "Dem sen' me fu' bring yu' dis talifone messige, en dem say, 'Hurry.'"

He handed me a note, which I opened hastily. It was from the rector of All Saints, Waccamaw, saying: "Mrs. S. of Rose Hill is dead, and will be buried at The Oaks at 3 o'clock to-day."

At once I sent a message to Bonaparte to have the red boat and Elihu ready to take me over to Waccamaw and I followed as soon as I could have the buckboard got. I felt such a pang that I had allowed myself to be discouraged, and not gone over to see Mrs. S. when the impulse seized me to do so two weeks ago, and now her remarkable personality was gone.

When I got to the plantation I found that Elihu, with all the other men on the place, was cutting rice in No. 8, the field they had rented from me, which is some distance away, and it was late before I could get Elihu. When he finally appeared he seemed much indisposed for any exertion; said he was worn out; could not possibly row to Waccamaw, so I concluded I would have to give up going, and I said with a sigh: "I am sorry I cannot go. Mrs. S. was papa's cousin. I have never seen her since he took me to visit her when I was a child, and I would have liked to attend her funeral, to do the last honors to her."

I was just thinking aloud as I often do. The change was instantaneous, as though I had used a magic word. Elihu exclaimed, "Ole Mausa cousin fun'ral! Miss, yu's got to go, en I's got to tek yu; I kyan't trus' no one else for tek yu wanderin' trou' dem crick; I bleege to tek yu," and Bonaparte echoed like a dignified Greek chorus, "Ole Mausa cousin fun'ral. Yu got to go," repeated several times, as though life, death, eternity, nothing could count under such circumstances.

The red boat was rapidly got out. It was leaking badly, and they made a little platform of boards where I was to sit, to keep my feet dry. As I finally got settled in the boat I asked Bonaparte the time and found it was 3, the hour I should have been at The Oaks. However, being in the boat, having overcome so many obstacles, I determined not to be daunted by the lateness of the hour — also the clouds had gathered heavily and a sprinkle of rain was falling. I borrowed Bonaparte's competent looking silver watch, for no watch that I have ever owned could stand my strenuous life for more than three months, so that I have to do without one.

I asked Elihu when we would be back, as we glided through the first canal. He answered: "Not till long after dark." I remonstrated, "Why it surely is not further than Waverly," but he answered, "'E mos' twice es far." I felt a little dismayed, but we kept on through endless windings past Cherokee Canal, then Long Creek, then the Thoroughfare, and at last into the broad, beautiful Waccamaw. When we reached The Oaks, the competent watch pointed to five.

It was a deserted tropical looking landing, with no living thing in sight, only the ruins of some houses. I got out and followed the road until I saw across a field at some distance several vehicles. I walked toward them and found the procession had just arrived and were carrying the coffin into the graveyard, a private one, with high brick walls and many monuments of past generations. Among these is one to Gov. Joseph Alston (Mrs. S.'s uncle), the husband of the beautiful and ill-fated Theodosia Burr; also their son, Aaron Burr Alston, whose death almost broke the mother's heart.

The sacred spot looked very solemn with its heavy live oak shadows in the darkening afternoon.

Mrs. S. was 86, I believe, and a saint upon earth. Since the death of her twin sister some years ago she had lived alone on her plantation, doing all the good and kindness possible to the people around her, who had formerly belonged to her. For the past two years one of her nieces had been with her always and was with her on the plantation when she passed peacefully away. There were forty or fifty of her people who had followed her to the grave and stood near showing every sign of grief.

The beautiful service of our Church was read, and we sang "Rock of Ages," in which the negroes joined with great fervor, weeping softly and swaying in rhythm to the music. I had only a few minutes to see the family, as they had a long drive I knew, and I a long row as well as drive. They

The sacred spot with its heavy live oak shadows.

kindly insisted on my being driven down to the boat, and I started home.

After that sudden gleam of sunshine the mist had settled down again over everything. How Elihu found his way through all those creeks was a marvel. I hoped when the moon appeared it would be clear; but it was covered with white clouds, which made a soft whiteness which was most confusing, and it looked to me always as though we were in an oval lake, from which there was no egress. It was most beautiful and mysterious, and I greatly enjoyed it. When I got home at 9 o'clock I was very tired and stiff from the four hours in the boat, for I had forgotten to take a cushion, but I was very glad I had gone.

Am going away from home for two weeks and always feel nervous and anxious as to how things will go on during my absence. I hate to leave Chloe so sick and suffering, for she misses me greatly and has only Dab to depend on in the yard. Besides, my neighbor has lost three fine horses in the last three days with blind staggers, and I feel as if I may find all mine dead when I get back.

CHAPTER VIII

PEACEVILLE, November 3.

I DROVE up from Gregory alone yesterday, reaching the village just at dusk. I thought with delight of the peace and quiet of the pineland settlement after the distress and indignation which I had felt since I left it.

Dab ran to open the gate, and Chloe had a nice supper ready for me, but I felt something in the air that made me lose the restful feeling, and as soon as I had finished my dainty little meal and Dab had cleared away things Chloe came in arrayed in the spotless white apron and kerchief which I dread, for they mean something serious.

After a few trivial efforts on my part to keep on the surface, for I was so tired and did so wish to float a little while, Chloe cleared her throat and began: —

"Miss Pashuns, ma'am, I cry studdy frum Tuesday night till now."

"My dear Chloe," I exclaimed, really alarmed, "how distressing! Have you lost some of your family? Not your grandson, I trust!"

"No, ma'am, I wouldn't a-cried es much fu' dat! No, Miss Pashuns, dis is wuss! I cry en I cry en I cry."

"For Heaven's sake, tell me, Chloe, what has happened!"

"Miss Pashuns, I know it would dustress you so dat I wouldn't tell you till you dun eat yo' suppa, 'case I say maybe yu might faint 'way w'en yu hear."

"Oh, Chloe," I cried, "I will faint away now if you don't get on and tell me what has happened."

"Miss Pashuns, Dab shot Mr. 'Apa's dog!"

"Impossible, Chloe! When?"

Then followed a long narrative which I did not altogether understand, but she said: "Didn't bin fu my gone to see Mr. 'Apa an' cry an' baig um to wait till yu cum back Dab wud 'a' bin on de chain gang by now, fu Mr. 'Apa bin dun indict um, but I baig um for hab de case put off till yu cum back. Happen so I hear 'bout um een time."

I called for the lantern and went off at once to Mr. H. Chloe begged me not to go out alone into the night and said she thought Mr. H. and all the family would be in bed, but I felt I must know the worst before I went to bed.

When I knocked the door opened on a pretty picture of home, a beautiful young mother leaning over a six weeks' baby in a big rocking-chair used as a cradle and four boys sitting around the fire. I begged Mr. H. to speak with me a few minutes in the piazza, as I thought it best not to discuss the matter before all those boys, though it was certainly much more comfortable inside than out, for it was sharply cool. As soon as the door closed I exclaimed: —

"Mr. H., I am too distressed to hear that Dab has shot your dog! I cannot tell you how sorry I am! Is it dead?"

"No, Miss Pennington, he never shot the dog at all, and I don't think he meant to shoot him either. This is the way it happened: —

"I had been out on a deer hunt and was coming in a little after dark, my hound dog running ahead of me. I heard him bark at something when I got near my house, but it was too dark for me to see what it was till I heard the report of a pistol and saw the flash and the ball dropped near by me.

"When I got near enough to see it was Dab; he said the dog jumped at him and tore his pants and he shot to scare the dog, but he said he didn't shoot at him and I don't believe he did, but I indicted him because it was a very wrong thing for him to shoot right in the road that way;

he might have shot some one; indeed he came mighty near hitting me, and he had no business shooting a pistol anyway; it's against the law.

"So I indicted him, but I told the Magistrate not to proceed till you got home."

I thanked him very much for his consideration and after making a little visit to the cosey party inside I went home. I asked him what I had better do, as I had not the faintest idea, never having had anything to do in law-courts.

He advised me to go and see the Magistrate and said that if any compromise could be made he would not push the case. He knew the punishment was a fine of $50 or thirty days on the chain gang.

I was quite overcome by his kindness and magnanimity in the matter and tried to say so, but by this time I was so exhausted that I fear I was not eloquent, to say the least.

This morning I interviewed Dab on the subject, speaking with all the force and wisdom I could. I cannot go to Judge H. until after to-morrow, for he will be busy with the election and have to go to Gregory, I believe — so I went to the plantation.

The quantity of peas gathered is most encouraging. I am quite delighted. I did not hope for half so many, and now the vines are being cut for hay with still a great many pods on them. It has not been cold enough yet to blast them.

The colts are growing finely and came running up as soon as they saw me. All the creatures, horses, cows, pigs, and sheep, are well, and I derived my usual refreshment and brightening by a few hours spent in God's good fresh air with the dumb things and the faithful trees, and came home quite cheerful and serene.

PEACEVILLE, November 4.

I had a sleepless, miserable night; the thought of Dab's going to the chain gang for firing a pistol was too distressing, and I saw no possibility of raising $50. It was a great comfort to me that Mr. H. did not believe there was any malice or premeditation in the act, but I knew very well there were others who would think differently. I have no idea what the mental attitude of the magistrate will be.

I talked with Dab a long time and told him that while I would protect him always from injustice I would never support him in defying the law in any way. I recalled to his mind what a long fight I had made with him about a pistol, how when he was a little fellow and somehow got hold of a pistol which looked as though it might be the Adam of all firearms, I had a procession formed, he leading, Rab following, and I bringing up the rear, down to the creek, and there I made him give me the weapon and I threw it as far as I could out into the water.

It is very hard to know whether Dab is impressed or not. I told him how miserable it would make me to see him go to the chain gang after all the trouble I had taken with him, hoping to see him grow up a respectable, honest man. I could not keep the tears from coming now and then, but Dab's black face was sphinx-like in its immobility. I told him I had not $50 to pay, and that all I could do would be to go to the magistrate and plead the fact that it was his first offence, and that he did not really understand what a serious matter it was to fire a pistol on the public highway.

November 5.

Drove up this morning to see Mr. H. about Dab's case. He lives at one of the old plantation homes, which has passed into new hands. I have never seen it since, and was quite moved in going there. It is a long, rambling house, a stone's throw from the Pee Dee River, surrounded by live oaks.

I left Dab to hold Ruth some distance from the house,

out of ear-shot. Mr. H. was holding court, so I could not see him at once, but his wife asked me in to the spick and span clean house and showed me her beautiful begonias, of which she asked me to accept cuttings, which I was pleased to do. When Mr. H. arrived I made my plea for Dab, and Mr. H. relieved my mind by saying as it was a first offence and Mr. A. was willing not to push it, he would try to arrange it "as light as possible."

He would send the Sheriff in a day or two for Dab, telling him to take the handcuffs along but not to use them unless Dab resisted arrest; but that if he came quietly, gave up the pistol and answered all questions frankly, he thought it could be arranged. He then set going a phonograph and treated me to "Rock of Ages" as sung by Trinity church choir, New York. *O tempora, o mores!*

I returned home, the horizon of my experience enlarged in an unexpected direction, wondering over the kindness of people generally, but very weary and worn with anxiety.

November 6.

The Sheriff came for Dab this morning while he was currying Ruth. Dab got his pistol from an old stump in the woods where he kept it hid, and gave it up and went quietly along. I do hope he will behave properly; he is always so respectful to me, but Chloe tells me he is not so to every one. If only the green-eyed monster could be eliminated, life would be easier.

10 P.M. I was so restless this morning that I had to write a few lines. Then I went to the plantation and about 3 o'clock I started back and met Dab on the road, returning, looking somehow solemn and made over. I was overjoyed to see him and tried to extract from him some account of his experience, but in vain.

His stammer was terrific, and all I heard was that Mr. H.

"I met Dab on the road."

read the law at length to him and impressed upon him that as it was his first offence and I had guaranteed that it would never occur again he would only have to pay costs, but that if ever again he carried a concealed weapon and shot it within fifty feet of a public highway it would go hard with him. I feel too thankful and relieved for anything.

November 8.

Sunday. A very beautiful day, and I am always influenced by that, so I begin to feel rested and more cheerful. We had a very pleasant service in the little chapel, and though both the alto and myself were hoarse the hymns were comforting.

My little Sunday-school children came in the afternoon and were very sweet. The lesson may not do them much good, but it does me a great deal.

CHEROKEE, November 11.

Got down to the plantation early, expecting to send Gibbie out with the ox wagon to move the heavy things. Found he had sent a message to say he was sick — a sad state of things, hands digging potatoes and no one to plough down the beds. I went to see Gibbie to see if he were really sick or only resting after his month's night hunting.

Cherokee steps.

I found him ill. I fear pneumonia. He is not strong enough for all that exposure at night. I refrained from saying "I told you so," but spoke very sympathizingly to him. His poor breath was so short; almost a pant.

I prescribed for him until we could get a doctor and had his wife lay a square of bacon skin sprinkled with turpentine over the side where the pain was. Then I sent for Green and told him he must take charge of the stable and cows until his brother should be out again, that I would let it go on his debt. He was very civil and said he would do his best, which was a great comfort. It has been a perfect day — bright, crisp, cool — and now that the effort is over it is delightful to be back in the large, pretty rooms of the Cherokee house, with everything pleasing to the eye within and without, and I feel very grateful to the Good Father. This is my mother's birthday and I would have liked to lay some flowers on her resting-place, but it was not possible, so I will go to-morrow.

Sunday, November 12.

Started very early and drove to Gregory with a great basket of beautiful roses. Was in time to attend service in the old church of Prince George — then I drove in to Woodstock and dined with my brother and was beguiled by the beautiful afternoon to stay rather late for my long drive alone.

The key to my front gate at Cherokee had been misplaced, and so I had to drive through the barn-yard way, which made it necessary to open four gates. It was quite dark by the time I reached the house and I was surprised to find it shut up and without a light. I tied Romola and went all around to try each door, but in vain; there was no one, and I concluded that Chloe had supposed I would spend the night at Woodstock and had gone in the woods to visit her sister.

Romola was weary and reluctant to start out again, but I determined the only thing to do was to go out to Bonaparte's house in the "street" and get him to come and break open a window for me to get in. I found it very hard to open all the gates again, but I got to the street safely and there I found Chloe. She said she had gone after Rab, who had stayed out late playing with the children, and she was afraid to stay alone at the yard, having just heard of little old Grace's death, my former cook, whom I was expecting to come back.

November 14.

Great excitement over the illness of my little grade Guernsey heifer Winnie. I had three people working on her from 11 till 4. To-night she seemed better.

Gertie is going to be married, so I have taken a new girl, to be here a week before Gertie leaves and learn something from her. When I asked Gertie what she wanted me to give her as a wedding present, she answered, after much bridling and what would have been blushing if her onyx hue had permitted, "Ef you could, ma'am, I'd like the 'sperience dress."

"And what do you wish for that dress?"

"Pearl gray cashmere, please, ma'am."

"That sounds very pretty and bridelike, Gertie, but I'm afraid it will be expensive."

"Yes, ma'am, 'e cos' 20 cent a ya'd to Gregory."

Much relieved to hear the price, I promised readily to get it. I have already provided all the materials and had the cake baked for her.

November 15.

Was particularly busy this morning when old Katie arrived. She comes about once in two weeks to ask for whatever she needs, and has done so for years, and I clothe and feed her in this way, giving her just what she asks for. Wonderful to say, this time she brought me a present of four eggs and I was quite touched. I gave her four quarts of rice, some grist, a small piece of bacon, and some milk, and after the politeness of the moment I returned to my work. Had not been fifteen minutes when old Louisa came with a little present of potatoes and a long appeal for sympathy and a letter from her daughter which she wanted me to answer for her. This took a long time. I addressed and stamped an envelope which I enclosed, so that Louisa will certainly get an answer.

Just returning to the putting down of the carpet, which I have to superintend most carefully in every detail, when a man came to ask to rent the estate farm on the sea-shore, so that I only got two carpets down, and finished those very late.

My good Chloe is in very bad humor, and things are difficult. She is furious at my having taken Gibbie to milk and cut wood and be about the yard, though she acknowledges that he does his work well, but he does not come of a family from which house servants used to be taken, and all the negroes resent his elevation to employment around the house, though he does not enter it except to bring in wood, which he does faithfully.

December 1.

Have been worrying a great deal lately about the taxes; they are nearly $200 and I do not know where the money is to come from. Mr. S., who has for several years visited this section buying rice, has written to me several times asking if I had any rough rice to sell. I did not answer from sheer *lachesse*. I hated to say that I had none. The little I have made this year I must keep for seed. To-day I drove to Gregory and met Mr. S. in the street, and he stopped me and asked if I had never received his letter. I answered just the truth, that I had no rice at all this year except seed rice, and only a little old rice left over, on which I had been feeding my stock, and I knew he did not want that, but he asked me to send him a sample of that at once, which I gladly promised to do.

I bought the pearl gray cashmere for Gertie's "'sperience" dress, a lovely looking soft stuff, truly only 20 cents a yard;

cotton, I suppose, but very pretty. Gertie was enchanted and said it was exactly right. Fashion is as exacting with them as with the highest social layer, and not to comply with what is just the last touch of elegance for a bride would be terrible to Gertie.

When I offered to give her the wedding dress she said it must be fine white lawn, and she would rather get it herself, as she knew where to get the finest, that means about 15 cents a yard, and she will have it made up in the latest fashion for 75 cents or $1 at the utmost.

December 7.

Took the long drive to Gregory again to receive and bring up a mare that has been sent me to keep for the winter. Having no one whom I can trust to go to Gregory without visiting the great moral institution, the Dispensary, I have to go myself and take Cable with me to lead the horse back. I have never taken him anywhere with me before, but he is a quiet, civil spoken negro, and comes of good family, and is not deaf like Gibbie, so I hired him to-day.

Met Mr. S. and he said he had written to offer $1 for one sample of rice and $1.05 for the other. I told him the letter had not reached me, but I would accept his offer. I tried not to let him see how surprised and delighted I was. After this I positively tread on air, for behold the tax problem solved, as I have nearly four hundred bushels.

To make my heart still lighter, Jim came to ask if he might speak to me, and he is anxious to come back. I think I discouraged him, unless his wife is willing to move into the country with him. He represents that he is getting very high wages, which he also represents that he certainly earns, for to use his own expression he "delivers a cow a day on his bicycle!" This marvellous statement means that he is working with a market, and delivers the supplies on a bicycle instead of a delivery wagon.

He says his health is wretched and he pines to come back to the country and to Ruth and Dandy and the other horses. I told him I could not possibly pay the wages he was getting, but he said he could save more with less wages in the country, which of course is true.

Altogether the day was a pleasant one in spite of fatigue and anxiety as to Nana's successful leading home. I proposed to Cable to ride her, but he seems very timid about horses, and though all battered and bruised from rough usage on the railroad she was full of spirit and anxious to have her own way.

I was afraid we might have trouble at the ferry, for Nana is a mountain horse and had never seen a ferry, accustomed to a bridge or a ford, and it would be decidedly awkward if she took a notion to ford the Black River, as it is (more or less) sixty feet deep at the ferry. I had behind the buckboard a bale of fruit trees from Berckman's done up in rye-straw, with the heads of rye left on. This Nana was so eager to get a good bite of, that she followed into the ferry without noticing where she was going, but when she found the flat in motion, she seemed frightened. I told Cable to let her eat all the rye she wanted and even the precious fruit trees, rather than have her begin to fight, and all went well. She stood quietly eating while we crossed.

December 8.

Not having yet succeeded in getting any one to patch the boiler to the threshing mill, I determined to go myself to Waverly and try to get some one. I sent Bonaparte to measure the cracks that I might take the measurements over with me, and told Gibbie to get the red boat ready for me. Gibbie, who is very deaf, did not hear "red" and I found him tugging at my white boat, which is up under the piazza, waiting to be caulked and painted. Fortunately I passed by and saw him, for it was an impossibility for him to move it unaided. I succeeded in making him understand that he was to bring the red boat from the barn-yard to the house-landing for me.

I went in and turned the eggs in the incubator, filled and trimmed the lamp, donned my boating outfit, and went to the landing. A long wait and then Gibbie appeared, looking hopeless. "Miss, I kyan' fin' Bony no ways, en 'e got de oah en de oahlock shet up een de bahn."

I expressed great impatience at this. Bonaparte should not lock up my oars. I always have kept them at the house, but poor Bonaparte knows his own race so well that he has an infinite distrust of them and locks up everything until it has become a mania.

Having suggested every possible place to find Bonaparte, at last I said: "Have you looked in the boiler?"

"No, ma'am," with a wondering look.

"Well, look in there at once."

He soon returned at a run to say that "Uncle Tinny and Uncle Bony were both in the b'iler" and wanted a lantern. This was sent, and after a prolonged pause they both appeared with the measurements of the cracks.

I patiently tried to understand Tinny's explanation as to where the holes were, but in vain. At last I said: "Anthony, you just get in the boat and go over to Waverly with me and you can explain to Captain Frank where the damage is, so that he will know what materials to send when he sends a man."

"But, Miss, I ain't fix fo' go. Ef yo' been tell me yo' wan' me fo' go to Wav'ly wid yo', I'd a dress maself, but I ain' fix; look a' me."

I looked, and truly the little gnome was an object — an old, tattered derby hat, with the mellow green tint of age, a very dirty new bright green and white plaid shirt, which only emphasized the extraordinary patchwork nether garments and coat, from the pocket of which conspicuously protruded a bottle.

With a grave face I assured the old man that he was quite decent and must go, and that as he was a fine paddler he could paddle while Gibbie rowed and we would go like a steam tug. This reconciled him to going in his working trim, and we started — I sitting in the bow with Tag, my nondescript terrier, Gibbie in the middle with oars, and the gnome at the stern paddling.

All this delay had consumed hours and the sky had darkened and it felt like rain. Chloe came to the wharf to beg me to wait, but I had wasted so much time and patience that I could not put it off.

I soon found it was a special mercy that I had caught old Tinny and made him come, for Gibbie proved a poor oarsman and the wind was against us and very high, so though we had the tide our progress was not rapid, and I was glad to have the old man, who knows all about boats. With the head wind Tag and I, high in the bow, were a great disadvantage. I longed for Elihu, for I would have felt safer with him.

To make things worse, when we got into the broad Waccamaw where the whitecaps were dancing, a steamer passed up the river, making such big waves that Tinny wanted to turn back without crossing, but I was not willing, as we were more than halfway to the mills, and to my surprise Gibbie supported me and we went on. Fortunately I had taken what the negroes call an " 'iler," a heavy rubber coat, to put over my knees. I had amused myself with pencil and pad, writing until the pad got too wet, for the water dashed in constantly. Poor Tag, straight up on his hind legs in the bow, looked out with dreary eyes, for at the best of times he hates water, and no doubt he said to himself that if he were a human he would have more sense than to leave a bright fire and comfortably carpeted room to be dashed and splashed in this way.

However, we reached the mill safely, and if only I had been successful in my errand, I would not mind, but Captain L. said he could not possibly spare any workman to send. This was a great blow, for I had written to him in June about it and he promised to send some one to repair the boiler, even naming a day when he would come. I do not know what to do now, for he knows all about such work and could tell me exactly what it was best to do, and I have such confidence in him. I did so wish to get the very little rice I have threshed out before Christmas. I will have to try to get a man from Gregory.

As I rowed up to the mills I came upon a flat heavily loaded with lightwood and recognized two of my men on it. I said, "Why, Billy, what are you doing here? Whose wood is that?"

"De my own."

Now I know why I have had so much trouble in getting my wood cut and sold. I had put Billy in charge and he has been steadily stealing my wood, he and his brother together, shipping it in a flat owned by their father, who is a gentleman of leisure living on his own land on a pension which he receives from the great Government of these United States.

Of course I will have to send Billy and Sol off the place after this discovery, for as Billy had been put in charge of the cutting and flatting of my wood and has so betrayed his trust, I cannot let him stay, but he will move to my neighbor's and continue no doubt to steal my wood, as his father's farm is very near my line. It is ten acres. Though every one on the place has known all summer what was going on, no one would give me the least hint of it, and I never would have known it if I had not made this trip to Waverly and come upon them in the act of unloading twelve cords of splendid lightwood. Of course it would be useless to take any legal steps when it is impossible to get testimony.

I got back very weary; it is astonishing how true the old saying is, "A cheerful heart goes all the day, while a sad one tires in a mile," and mine was very sad on the way home.

CHEROKEE, December 9.

Was very busy this morning writing letters to catch the mail which passes my gate at 11, when they came in to say that Annette wanted to see me "pa'tick'ler." I went out and said, rather shortly: "What can I do for you, Annette? This is a busy day and you must talk quickly." Annette twisted her hands together nervously once or twice, and then answered: "I came to baig you to make my will, ma'am."

"Why, Annette, do you feel ill?"

Her head went down and her apron came up, half covering her face, and she said:—

"No, ma'am, but I 'spects to be married again. I bin a fateful wife to St. Luke en' I bin a fateful widder to 'im f'r t'ree year; but now a very 'spectable man is co't me, en' I'se to be married next week, en' I wants to put all St. Luke proppity to 'e chillun, de house en' de fa'am between dem, en' de cow fer Annie, en' de two heffer between dem. De man I gwine marry got 'e own house en' fa'am, en' 'e seem to speak en' act very fair, but I wan' to lef' St. Luke chillun secure."

I was so delighted at this evidence of Annette's intelligence, knowledge of human nature, and loyalty to her dead husband's interest, that I forgot all about my important letters and drew up a most impressive document which I had her sign in the presence of three witnesses, being the disposal of real estate. She only "teched de pen," that is, put her hand on the end of the pen while I wrote her name, and she made a mark. When the will was satisfactorily executed she wanted me to keep it, but this I declined to do, advising her to give it into her mother's care, if she preferred not to keep it herself.

December 10.

Went to church, though there was a gale blowing and the trees looked very dangerous buckling and bending over the road. Ruth behaved well, though she did not like it. When I got back, bringing L. to dinner with me, I found Jim waiting to see me, having ridden up from Gregory on his bicycle. He said he wanted to come back, that his wife was not only willing but anxious for him to come, as she had no pleasure in his life in town, he was so ailing and worked so hard. He begged me to take him. The only thing he wanted to ask was that I would let him spend every Sunday in Gregory, for he sang in the choir in his church and didn't want to give up the music. I told him I would always do so if possible, but that there might be circumstances which would make it impossible to spare him on Sunday.

The smoke-house at Cherokee for meat curing.

He cannot come until he gets some one to take his place, but as he is coming I will put off the meat curing and sausage making until he comes, for my mother taught him the best way of doing all that, and it makes it so much easier than undertaking it with a green hand. Jim is to do almost everything, under our present agreement; but Gibbie is still to milk and to keep the stables clean and cut wood for the house.

December 11.

Lent Bonaparte the ox wagon and team to go to Gregory and lay in his household supplies. Sent a note to Billy P. — and his brother, who had been selling the lightwood for themselves instead of for me, telling them to leave the place with their families at once. Of course I should have taken the money for the wood, but I could not make up my mind to do that without some legal process, and as I could not get any witness to testify against them any legal process was impossible. If I had ordered them to give me one-half the money, quite possibly they might have done it; but they might have refused to do it and I would then have been powerless. I am very careful not to give any order which I cannot have obeyed.

December 12.

Billy and Sol came to beg me to let them stay until January, but I told them I had trusted them and they had betrayed my trust and must go at once. I hate to lose their wives, who are good workers, and their little children, who come to say catechism and sing hymns and have a stick of candy every Sunday afternoon. Sol's wife, Aphrodite, is such a specimen of health and maternal vigor that I delight to see her going to work with her procession of little ones behind her. The men themselves are strong, able-bodied workers, and I shall miss them; but once having begun to depredate upon me, nothing will stop them.

I find now that recently they have been living out of my vegetable garden, and the potato banks have been robbed and there are dark hints as to their guilt in that, too. I told them they must whip their rice out by hand at once so as to pay their rent, and take the rest with them. It is a sad state of things that one is unable to secure redress in any way for depredation, and so the only thing to do when a tenant goes wrong in this way is to send him off, so that unless one winks at evil deeds or condones offences, one will soon be without hands entirely.

December 13.

Yesterday, driving out, I saw a raft of very fine poplar logs being made down by the bridge in the creek, and this morning

Sol's wife, Aphrodite, is a specimen of maternal health and vigor.

I walked out to see whose logs they were. I have on my woodland across the creek some very beautiful poplars, some of them about three feet in diameter at the base. I have several times been offered a price for them, but have always answered: "When I am in need of bread I will sell those trees, but not before." Now I feared some one might have cut them, hence my desire to inquire about the raft.

I found Jack and Monday at work on the raft, which was composed of splendid pine as well as poplar logs. Both of

I saw a raft of very fine poplar logs being made.

these men had belonged to my father and now own farms and woodland of their own about two miles away. They assured me the logs all came from their own land and had been hauled with their own ox teams. I complimented them on the size and beauty of the poplar, and just at that moment Daniel, another one of our former people, now a prosperous landowner, came by in his canoe, and I took passage with him up the creek to my woodland, as I wanted to make sure that my poplars were still standing.

As I got out at the landing I offered to pay him, but he said: "Oh, no, Miss; you don't owe me a cent. I was just on my way home, and I'm glad to have de chance to do it for you."

I thanked him suitably for his pleasant feeling. At the landing there was a raft tied of very large logs. I asked Daniel whose it was and he said it belonged to Frank and Logan, who were cutting on my land. I was greatly shocked. Logan is the son of one of my father's most trusted servants who died a few years ago leaving eight sons and three daughters grown up and married. He was a first class engineer and blacksmith and could be counted upon always to do faithful, good work. His sons had most of them followed his trade after a fashion, and all of them had what is now called education (without, however, any training) and are smart men; but not one of them has the character, the thoroughness, the reliability, of the old man, who could neither read nor write, but who had been trained to do one thing as well as it could be done.

The sons have, one by one, left my service to go where there was more demand for their work and more pay, but a kindly feeling has remained between us. They are all prosperous, living on farms of their own.

Some months ago Logan brought Frank, a stranger to me, to ask to handle the fallen pine trees on my land and give me one-third of the proceeds. They said they had fine ox teams and each had a logging cart and were fixed for the business. After thinking over it a while I consented, for I had been over the land and knew that there were many fine large trees blown down by the storm which would only rot on the ground if I refused, for I had no reliable hands to get them to market myself. I made them sign a paper saying they were to cut no trees, only to take the prostrate pine, and was quite pleased when the arrangement was completed.

The results had disappointed me, being much less than I had supposed they would be. Every now and then they brought me $8, with account of a raft in Captain L.'s hand-writing showing the amount of my third, and I had been suspecting that they were carrying many rafts to Gregory and selling them on their own account, not giving me the third, but I did not see how I could find out the truth. They had come to me in the winter to ask permission to "dead" some cypress. This means to ring the cypress so as to kill it, otherwise it cannot be handled; it will not float if green.

Cypress trees.

I refused to give permission for this, and a short time afterward they asked to be allowed to cut some poplar. This I forbade with horror, and they went away. Now the sight of this raft made me understand of what treachery they had been guilty toward my trees.

I counted the logs — twenty pine, four cypress, and two poplar; then I walked out into the woods and soon came upon Logan with his team hauling a splendid log and Frank not far behind with another. Just for a moment, as I stood waiting for them to come up, it flashed through my mind what a rash thing perhaps I was doing, as both of these men are rather ugly tempered. I had sent Daniel off with his boat, thinking some one might come along the creek by the time I was ready to go back. No one at the house had the least idea where I was, for I had not intended making this extended trip when I left.

When the men came up I taxed them with having done what I had expressly forbidden them to do. At first they were disposed to be rude and answered roughly, but I went on very quietly, using all the self-control possible, to tell them that they had violated their contract and put themselves entirely in my power; that I needed no witnesses, for my own eyes had seen what they were doing. Gradually their whole manner changed. One hat went off and then the other and Logan came a step or two nearer, and with a most dramatic air of humility and penitence said: —

"Miss, you right; en we cry guilty, guilty! We own um, we's guilty, en you know, miss, w'en a man stan' 'e trial een de co't, en dat man cry guilty, de jedge don' put de law so heaby to um. We dun wrong, miss; we egkno'ledge we sin, en we pleads yo' mussy!"

I was completely taken aback. I was prepared for anything but this, and I had no idea what to say in my surprise. While I considered they stood with bowed heads, eyes fixed on the ground, and every air of complete surrender. I was disarmed, and of course did not follow up my victory as I should. I gave them a little discourse on judgment and mercy and on the awful sin of deceiving and taking advantage of one who had trusted one. Then I told them they could take the timber they had cut and hauled, to market and give me half instead of one-third, and that after selling these logs they must not touch a stick of timber of any kind again.

With expressions of profound thanks they led the way down to the swamp as I told them to do, and showed me all the trees they had cut. It was heartrending to see the havoc they had wrought, and which nothing could undo. It took away my breath almost for a time, and I felt almost as though I had been wrong not to proceed against these men and have them brought to justice. I knew perfectly I would get no money to speak of from them.

It is impossible for me to watch the woodland and swamp myself — if there is no one to see after my interest there it is indeed hopeless. Bonaparte used to do it, but now he seems to have been intimidated in some way, and will not undertake to see after it at all.

December 14.

It poured heavily all day. At 1 o'clock they came in to say the wagon had come for the cotton. Of course I could not send it in the rain, and I had to send the man back. He was very wet and cold and I gave him some potato pudding and milk, all that was ready. The gin is about twelve miles away and I had engaged them to send for my cotton to-day. It was folly to send in the rain. Still I suppose I will have to pay for it.

Chloe went to St. Cyprian's last night and had much to tell of the service and her approval of the sermon. She said Mr. G. was "a good preacher en preached de pure gospel." She told me she had walked back with old Anthony and that he praised the sermon and then told her of his dissatisfaction with his minister, a Baptist. She said : —

"Br'er Tinny say, him don' like de preacher dem got; say dem ax de man fo' preach out to Tolson village, en as him had to cum clean f'um Gregory ebrybody carry 'nuf money for t'row een, but w'en him beggin fo' preach dem fin' him preach politiks, en slur, en Latin, en dem 'ordn't t'row een dem money, en de man neber git but one dollar en a half f'um dat big crowd o' people."

"Well, Chloe, you will have to tell me what preaching 'slur' means."

"Miss Pashuns, dat mean him hol' up him perfesshun high, en him scandalize all dem oder Christianity, en dat mek dem feel shame en dem didn't like dat. Him bin a Babtist, yu see, en de chutch bin full o' Methodist."

"Oh, I understand now. That was very bad indeed; now tell me what does preaching politics mean?"

"Dat mean stid o' preach de gospel of de Lord, him bin a talk 'bout de State en de law, en de guberment, en 'e got dem all tangle up en dem mind."

"They certainly showed their sense, Chloe, when they objected to that, for they went to church for heavenly instruction; but tell me what preaching Latin meant."

Chloe seemed to be a little tired of my questions and to think me dull, which is not my ordinary trouble, but she explained: —

"Well, Miss Pashuns, yu kno' 'tain't ebrybody kin onderstan' Latin, en w'en dis man kum to a place wey him hab nuthin' sensible fer said, him sta'at fu' ramble een a kin' o' gibbish en nobody c'udn't onderstan', en de man's self c'udn't onderstan'. Bre'r Tinny say you c'ud look een 'e eye en see him jes' bin'a wander. Him didn't hab nuthin' fu' say; so him didn't t'row een him money, en say w'en he yeddy * Animus Brown is fu' preech him stay home."

I was quite amazed that little old man Tinny should have such power of discernment, and also such apt terms to describe and size up his preacher, and I was truly thankful he recognized the difference in Mr. G.'s doctrine and methods.

The darkies have a wonderfully keen insight into character. It is almost as if by instinct they know the genuine article from the imitation, the gilt from the gold. When you look at Anthony you would not think he knew anything more than a sparrow sees with its beady black eyes. He is very dried up and little, with those very same beady eyes. I think a great deal of the old man; he makes me a present of a huge pumpkin every year, and after many efforts to find out what he would like in return I make my present.

Sometimes I am baffled as to what he would like and give him money. If I do this, the very next day he hangs his

* "Yedde" means to hear in real gullah, which some of the old darkies still use.

shoes on a stick over his shoulder and walks down to Gregory, fourteen miles away, invests his cash in firewater and walks back home, all with a little shuffling gait which makes it hard to believe he could walk twenty-eight miles a day.

December 18.

This afternoon Gibbie came to say that his mother was very low and so he would not be able to milk, so I took Goliah to the cow stable to help Bonaparte milk and then to put up the horses. Many little negroes of Goliah's size are good milkers, but he has no skill in that line at all, though he is remarkably clever and useful with horses.

December 19.

Both yesterday and to-day I got up very early and went out to the stable to help Bonaparte. It is very provoking of Gibbie to absent himself in this way, for I find he is not waiting on his mother, who has her husband and three other sons and their wives devoted in their services, while Gibbie is just idling along the roads.

December 20.

A perfect day, the air warm and balmy. On my way home from church heard of Eva's death. She was a simple, faithful soul, always diligent, working hard in her large field around her house and giving freely of the produce to her five sons, four of whom have families, but none of them has inherited her working, faithful nature. I will miss her greatly.

I had a good attendance of darkies at Sunday-school this afternoon. I was so pleased to see the children all so clean and nicely dressed, and they behaved so well. There were fourteen girls and fifteen boys, most of them between 10 and 14 years of age. After they have gone over the Creed, the Lord's Prayer, and the Ten Commandments several times, with explanatory remarks from me, they repeat after me a hymn, this time: "While Shepherds Watched

Their Flocks by Night," preparatory to Christmas. Then I go in to the piano and have the girls in the room, while the boys stand by the window; and they all sing *à faire peur*.

They enjoy it so that their whole strength is put into sound. In vain I listen for the sweet voices I have heard in times past

She was a simple, faithful soul — always diligent.

— this is all volume without sweetness — and I fear I will crack my own throat in my efforts to guide the volume aright. "Jesus, Lover of My Soul," they know pretty well, also "Onward, Christian Soldiers."

After four hymns they stand in order of size in the piazza and I hand around two pounds of candy, which just gives each child a stick, and they depart. But to-day little fellows shot out from the row and four with much serious unwrapping of handkerchiefs handed me each an egg. I was much surprised and thanked them with effusion.

They come every Sunday before I have finished my dinner, greatly to Don's indignation; any arrival at meal-times is displeasing to him, and for fear he will frighten the children I have him chained as soon as I come from church on Sunday. These children are all grandchildren of those who belonged to my father.

December 21.

Bagging rough rice in the barn all day. It is very cold and dusty. I have most unexpectedly sold this rice for a dollar a bushel, and instead of being full of thankfulness, my poor human nature is lamenting over the 600 bushels which I have fed to my creatures all summer, and let the hands have whenever they wanted it for forty cents a bushel, and thinking how rich I would be if I had it here now. I cannot get the rice from this year's crop threshed, little as it is, because it seems impossible to get any one to work the boiler.

December 24.

Very busy putting up a parcel to send to Dab's little brother Rab who is, I hope, being made over into a very good boy by the worthy Jenkins. The parcel contained only a suit of clothes, caps, suspenders, and necessary underclothing, but I wanted it to reach him on Christmas. To my intense regret I could not put even a nickel in the pocket. I generally put a quarter and I know he will search every corner.

The mail brought packages with loving offerings from my dear ones. I had not the heart to accept the many invitations I had to spend Christmas, and so I am alone and have time to realize the one great Christmas gift made to our humanity once for all time. This evening I sent by Gibbie a little package of good things to each darky child on the place.

Christmas Day.

I sat up until 1 o'clock last night rummaging through my possessions to find presents for the servants. I cannot bear

Winnowing house for preparation of seed rice.

to have nothing for them, but my dear father's constant injunction, "Be just before you are generous," is indelibly impressed upon me. I owe money to several and so I have not been willing to spend even a quarter on Christmas preparations.

All the grown servants have gone to the "setting up," which is one of the strongest articles of their creed and is very impressive, I think — the feeling that they must not be found in their beds on this mysterious night when the King of the world was born and laid in a manger. A feeble old woman with whom I remonstrated, telling her she was not strong enough to sit up all night, turned on me in indignation, saying: "Miss, yo' t'ink I 'ood let de Lawd ketch me in baid to-night w'en de bery cow fall on dey knee! No, ma'am, dis night is fer pray, en shout, en rej'ice."

My packages yesterday contained six boxes of candy, four of them the most delicious home-made nut candy of different kinds. I had two pounds of common stick candy in the house, and after getting up some ancient silk things, I found five boxes to fill, one for each servant; the stick candy in the bottom, and some of my delicious things out of my recent presents to fill up.

I was so in earnest searching in the garret for empty boxes by the light of a dim lantern that I did not notice what labels they bore until I had filled the five and put a Christmas card on top of each and put the covers on. Then I laughed until I cried. The largest, which I had put particularly nice things in, was labelled "Finest mourning paper" and had great black bands all around. The next was labelled "Best carbolic soap," and the others were also soap boxes. It was too late to take out all the carefully arranged contents and begin over, so I tied them up with ribbon and put two apples on each so that they would be on hand when I heard the call: "Merry Christmas!" at my door in the early morning.

In the olden time there used to be such crowds coming in to the upstairs hall to wish the Merry Christmas, and one must have a gift for each. Long after the war they kept it up, and I used to have a hamper of little gifts all wrapped to pitch out of the door as I heard each voice. Now I had only Chloe, Dab, Betty, Bonaparte, and Gibbie to provide for. I put up little packages for old Katie and all the old darkies who come to the yard to wish us Merry Christmas and bring an egg or two and receive their Christmas.

This is a survival of the past, when every negro on the plantation came soon after daylight Christmas morning, to give their good wishes and to receive substantial gifts themselves. They always had three days of entire holiday, during which they amused themselves, always ending the day by two hours' dancing on the piazza of the "big house" to the music of fiddle, tambourine, bones, drum, and sticks. My father sent off young lads to learn to play the violin every year, so that there were always one or two capable of leading.

The way in which they mark time with the sticks has always been a wonder to me. They beat them in syncopated time, the accent always being on the second beat. I have tried in vain to get the motion, and yet very little children do it in perfect time.

I drove to church thinking of all the nice things I would like to be carrying to my friends in the dear little settlement who all sent me some charming token of affection and goodwill. Only three or four assembled and the holly filled font was the only sign of the great festival. Our organist was not there, so that I knew I would have to "raise" the hymn — that means stand up in your pew and sing it without accompaniment.

What was my dismay when "Shout the Glad Tidings, Exultingly Sing," was given out. There is but one tune that I ever heard to it, and that is most elaborate. However, it is the forlorn hope that rouses and appeals to me. I rose to my feet and the occasion, and the glad tidings were shouted most enthusiastically by one feeble voice. Only at the chorus Miss Pandora gave the support of her voice. It is pleasant to remember that the Good Maker of all, does not have to listen; he looks within and sees the spirit which impels those inadequate sounds.

I came away from the simple service in high spirits, all my depression and discouragement floated upward in the quavering shouts of glad tidings.

December 26.

Rode out to the post-office on horseback and enjoyed it immensely. Got a charming book there. I have had so many dear little presents, the most valuable being a pair of driving gloves, which have delighted me.

December 27.

Started to church this morning with Ruth in the buckboard and found her dead lame! Had to turn and have Dab get Romola out of the field and put her in.

I am worried because the stable door is off its hinges, and it is strange Gibbie should not have reported it yesterday, for Bonaparte could have fixed it in five minutes. I used Ruth Friday and she was quite well. I fear she has been ridden at night and put her foot in a hole.

December 28.

J. and L. came Sunday evening and spent the night. Yesterday morning L., who has made quite a specialty of animal diseases, examined Ruth's leg and foot. He said it was the hock, and he only hoped it might not prove spavin — said it must be bathed twice a day with hot mullein tea and then rubbed dry.

I begged him to tell Gibbie exactly what to do and how to do it, as it would impress him more, coming from him. So he gave him most careful instructions about it. This morning at ten, when I was rushing with some letters to catch the mail at the avenue gate, I saw Ruth standing near the back door with the long strong reins, which Jim made to drive the colt, wrapped 'round and 'round a walnut tree. I was so provoked that I forgot the mail and addressed Gibbie, who was rubbing Ruth's leg and asked why he had put the head-stall and reins which had been made for the colt on Ruth. He answered: —

"Jus' so. I had dem on Alcyone driving about in. W'en I stop I put 'em on Root."

"Where is Ruth's halter?" I supposed something had happened to it, but not at all; it was in the stable.

As I saw the large pool of water on the ground and Ruth's leg rubbed quite dry I grew milder in my words and simply sent him to get the halter and put it on. Then I went on with the letters. As I came back I went into the pantry where the breakfast things were not yet washed. Again I exploded.

"Why were the cups and silver not yet washed?"

"Gibbie cum een en say he had o'ders to git all de hot water fer Root fut, so 'e hempty both kittle, en we had to put on water fresh f'r we, en it ain't hot 'nuff yet."

I simply had to leave without saying a word.

Later Chloe sought me when I had just settled myself at my desk to write, and said: —

"Patty en Dab en me all bin a eat."

"Miss Pashuns, all dat hot water Gibbie tek out de two kittle out de kitchen, 'e neber put a drop on Root fut!"

"My, Chloe! what did he do with it, then?"

"'E po' um right out down on de groun' long side a Root een a puddle, en 'e neber so much as tech Root fut wid a wet clot'. 'E rub um wid a dry rag."

I cast my mind back and remembered how very dry Ruth's leg was, and how pleased I was that Gibbie should have rubbed her so well; but still I could scarcely believe that he had poured all that hot water on the ground. While I was considering, Chloe went on: —

" 'Tain't me one see um. Patty en Dab en me all bin a eat we break-us en we set down right dey, en luk at Gibbie when 'e pou' out de watah 'pun de groun'."

What a farce to try to have anything done at all! I did not say a word to Gibbie about this. He would simply swear it was not so and there would be a state of warfare in the yard.

I sent for him that afternoon and told him to fill a big pot in the yard with water, get some mullein leaves and put in it, make a fire under the pot the first thing in the morning, and after cleaning his horses to bring Ruth out and let me know, as I wished to bathe her myself.

CHEROKEE, December 29.

Jim is not coming back; his month with me is up, and he has work in Gregory. Now the question is either to give up all the progress which has been made in Marietta's training or to drive her myself.

I have always been afraid of a skeleton road cart, and I confess I dread driving in one. I asked Jim before he went one day if he could not try her in the old buckboard, which is very light. He said by no means, that she wheeled square round at any new thing she saw, and would break the shafts at once.

To-day I had Romola put in the road cart for me to try it, and drove eight miles. The seat is just an unusually hard board, and I knew that the least thing would make me pitch out. When I got back I called Bonaparte and had him take off the board and put some strong wide pieces of leather across and then tack a sheepskin on top, and I will try it to-morrow.

Then I told him I wanted a small seat secured to the axle at the back so that Gibbie could sit there. Bonaparte indicated that ordinarily he found my plans intelligent, but that in this instance he failed to see any sign of common sense. It was all in civil, even courtly, language, but the meaning was plain. I was not daunted. I said: —

"I cannot go out alone with that colt. I must have Gibbie at hand, and the only way to take him is to rig up such a seat, and I trust to your cleverness and skill to do it."

I got a very strong chair of white oak and had him saw off the back. "Now this is what I want you to use, and I want you to put it here," I said. Then I left him.

December 30.

This morning when I went to look at the progress of the little shelf behind the road cart, I found Bonaparte working with enthusiasm. The idea had suddenly taken him, but Gibbie was looking on with a face of woe, muttering steadily "risk my life — got wife and chillun — brek me neck" — I could only hear a word here and there.

"Gibbie," I said, "how many times were you thrown out of the road cart when Jim and you went with Marietta?"

"Only fo' time."

"Where did you sit?"

"Jim en me set on de seat; each one had a fut on de step so we could jump quick."

"Did Jim jump out?"

"Jim jump out 'eself ebery time we meet buggy, dat colt wunt pass a buggy, en de las' wud Jim say to me was, 'Fo' King sake Gib neber let um meet a w'ite hoss, kase 'e'll bruk up eberyt'ing.'"

"Well," I said, "now you will be perfectly safe behind here, for when anything happens you can step off without any trouble."

But Gibbie continued to grumble and mutter. As soon as the extraordinary little perch was adjusted I made him put Romola in the cart and took him behind for a six-mile drive. He nearly refused to go, but I kept my eye on him, and we started.

We had not driven out of the front gate before I heard sounds of satisfaction from behind — little grunts only at first, but at last he burst into speech. "My law, Miss Pashuns, you hav a good idea w'en yu fix dis seat! I too cumfutable! Jes' es easy es if I bin home een me rockin' chai'. Dis' de t'ing fo' me." I was greatly relieved, for as Gibbie has been going with Jim every day, with Marietta, it is important he should go along with me. One change at a time is enough and I cannot let him drive, because he has such a heavy hand, accustomed to handle oxen, but I could not take him if he was afraid or unwilling.

I was equally delighted with my seat, for the sheepskin made all the difference; one could sit home as on a saddle. I did not think the cart balanced just as I wished, however, and when I got home I told Bonaparte to get a heavy piece of iron from the old mill and fasten it where the dash-board is in a buggy; this cart has none. He did this and I got in and made Gibbie get behind while Bonaparte steadied the shafts and they stood level without his holding them. Then I was satisfied. Everything is ready now and to-morrow I will drive Marietta.

All the neighbors are making an outcry about it and my dear friend Miss N. to-day said all that could possibly be said to deter me, but I cannot see it as they do. My taxes are $100; they are due now. If nothing turns up I must sell something to pay them. Last year I sold a colt for that purpose. Now Marietta unbroken would not be salable, but broken she would bring a good price.

It will be a heartbreaking business to part with her. She is exactly like her mother and they would be a delightful pair, but I must try and get her broken if I can. She has made good progress in a month, I think, for she was not even halter broken the first of December.

December 31.

Started with Marietta at 12:25 to-day and drove eight miles, getting back at 1:35. It was truly exciting, but she went wonderfully. All the way to Peaceville we were so fortunate as not to meet a vehicle on the road. Coming back a buggy turned into the narrow road ahead. I waved to the man to turn back, but he did not understand, for I would not speak to let Marietta know I was telling him to go back; but as soon as she saw him she made herself immense, and began to trumpet like an elephant, standing stock still.

The man needed no suggestion after seeing and hearing her, and rapidly got out and lifted his buggy around and fled into another road. After a while she quieted down and we went on. It is a great pity to have such a road; it is barely wide enough for two vehicles to pass and there is a deep ditch on each side.

We had gone about a quarter of a mile after this and she had steadied into a quiet trot, when two dogs, one white and one black, dashed from a house about 500 yards from the road, and rushed toward us, barking furiously. This was too much. She started at a full run and all my effort was directed to keeping her in the road, for those deep ditches so near on each side were a terror.

I talked to her as I put out my whole strength on the reins. I felt I could not stand it much longer, my arms were giving way and I wondered whether Gibbie was thrown off and what would be the end, when she slacked her speed and finally came down into a trot. Then I called to Gibbie.

He is stone deaf, which makes a difficulty, and I was too shaken and stiff to be able to turn around to look; but when at last he heard he answered with cheerful equanimity. Having once given me his faith, Gibbie did not appear to have the least anxiety.

My heart was filled with thanksgiving as I stepped down from the funny little cart at the stable door, Marietta dripping with sweat and blood streaming from each side of her mouth, but turning around to see what I had for her with a look of affection. I always gave Ruth a lump of sugar when she had been good, but this poor dear little hard times thing won't take sugar nor apple nor carrot — no, nothing but an ear of corn will she take.

This is the last night of the old year. For the first time since the tragedy I felt myself drawn to the piano, and I played Chopin's funeral march over and over, with its wonderful wail of sorrow, and then Beethoven's funeral march on the death of a hero. Such a contrast! No wail here. Rather "Gloria Victor": —

> O Death, where is thy sting?
> O Grave, where thy victory?

The old year is dead. God grant us grace in the new.

CHAPTER IX

January 1.

THE new year ought to fill one with bright anticipations and hopes, but somehow I am so weighed down by realities, in the shape of bills and accounts which should be paid and for which I see no wherewithal, that my horizon seems dark and cloud-capped. I try to keep myself hard at work, as that is the only way to get rid of anxiety.

I am having wool washed to make a mattress, as I need a nice single mattress, and the only way to get it at this moment is to make it. Chloe and Patty are to wash it to-morrow.

January 2.

Drove Marietta this morning for the second time. Jim always walked behind driving with long reins, while Gibbie led her as far as the front gate, so I followed his example and drove from behind until we got into the public road, when I got in. She fought a little, but went beautifully when once we were started.

I wanted to go to the post-office in Peaceville but did not wish to stop there, as we did that the last time, and with a colt it is so important not to let it form a habit of stopping at any one place. So I drove all the way up the village to the last house, and turning there came back to the post-office. Alack and alas! it was closed. If only I had stopped on the way up I would have got my mail, and I was hoping for a valuable letter. While I stopped talking, asking if it would not be possible to have the office opened for a moment, one of my dear Sunday-school boys galloped by on horseback, followed by his black dog.

Marietta just made up her mind to get rid of all impediments and pursue them, especially the dog. She reared, she plunged, she bucked, she whirled, she stood so long on her hind legs pawing the air that I thought she must fall back on me. Gibbie, however, held on to her manfully, although nearly lifted from the ground. Mr. R. was so excited that he jumped the high paling fence to come to my assistance, but there was nothing he could do. However, I was glad of his suggestion, made in his deliberate way: "I would turn her head the other way," which advice I gladly followed and drove rapidly up the village for the second time and on in that direction until she was somewhat quiet and then turned homeward, trusting Fred and his black dog had gone a long distance and would not return until I was safely at home.

One dangerous spot after another was passed and I began to breathe more freely when, as I reached the Clay gully, I saw in the distance the galloping horse and frisking black dog approaching. My heart was in my mouth, but I make it a rule never to call out in an alarmed tone, as a horse is so sensitive to the driver's feelings. I had taken her around a little side cut she was unaccustomed to so that she was so busy examining every root and stump that she did not see the approaching party. To my great relief Fred saw us, and with wonderful presence of mind called his dog, which had nearly reached us, and rode rapidly off in another direction. I was very thankful, and greatly pleased at the boy's prompt thought and action.

Got home without further trouble and did not give poor little Marietta the reward she was expecting — two quarts of oats. Gibbie was indignant at this and proceeded to argue with me, but I was firm and told him Marietta would understand perfectly.

January 6.

Have had the great privilege and pleasure of having our Bishop as my guest, on his pastoral visit to our struggling little parish. The Bishop's visit is always a season of uplift and thankfulness.

January 7.

Drove Marietta to-day, and though she was nervous at first and it was hard for me to get in the road cart, she soon quieted down and went eight miles without any excitement, so that I had the pleasure of giving her the two quarts of oats mixed with soda and hot water which is the reward of merit.

Friday, January 8.

This morning I told Gibbie that we would drive down the road, as we have been up so often, because the bridge a short distance below has been undergoing repairs. Marietta went very quietly until we got out of the gate and turned her head down the road and I got in — then she wheeled sharp around and reared until I thought she must fall back — she plunged, she squatted until she broke up the harness entirely. Gibbie lost his nerve and instead of holding her by the bit, as he did the last time she fought, he held the end of a six foot halter, so that he had no power over her and was in danger of being pawed.

I held on to the reins, fortunately. She turned herself around in the shafts, having broken girth and crupper, until she faced me, and as I kept my tight grip on the reins she was nearly choked. Purposely I pulled tighter and tighter, and when she found herself entirely tied up in the harness and choking she was quiet and stood without moving while Gibbie and Bonaparte took off the remnants of the harness. Fortunately the head-stall and reins were strong and held. I found there was no hope of putting her back in the cart, as it would take days to patch up the harness, so I told Gibbie I would drive her down the road without any vehicle, he leading and I holding the reins behind. We had a great deal of trouble to get her started down the road, but she went after a while quietly enough until we came to the bridge, where she made a tremendous fight. When I was worn out with her wheeling and fighting I gave the lines to Gibbie and told him to stand perfectly still, not make any effort to get her over, but if she started to go, to follow her. Then I went across and stood a short distance from the bridge and willed her to come over, putting all my strength into the will. She put one foot slowly forward and then the other, apparently with the greatest reluctance, but once started she came straight to me, and then I took the lines and drove her three miles. She was just as quiet and docile as though she had never fought. She walked so rapidly, however, dragging me along at a most unusual pace for me, that I was completely exhausted when we got home.

January 8.

Started on mattress about 10 o'clock and worked steadily until I finished it at midnight. I made the tick on the machine just after breakfast and then had Bonaparte make me a frame just the size of the spring I wanted the mattress to fit. This was not finished until 10 and I was very much afraid I would not be able to finish, but I did by working, with only half an hour for dinner. I get so interested in anything I am doing, it does not matter what it is, for the moment, it is the most engrossing occupation in the world. The wool was beautifully washed, which made it pleasant.

When it came to sticking a needle a foot long through the mattress and tying with twine I had to get Jim's willing and efficient help, but that was not until after 9 to-night. I am so exhilarated by the success of my work that I am neither tired nor sleepy and have to make myself stop working and go to bed, when I hope to sleep serenely "clothed in the light of high duties done."

January 9.

Sewed nearly all day, which is a rare treat to me. The wood we are using burns out so fast, that I have been urging the men to cut enough logs from the live oaks (which I have at last got sawed down), to give each fireplace a back log; that makes such a difference in the permanence and heat of the fire. Joe Keit said the wood was too hard, might as well try to cut iron, and that it would take all day to cut one log, making it very dear wood. I was provoked, but never having sawed any wood at all, I did not know whether what he said was true or not — that always worries me — so I put down my sewing and got the big saw about $4\frac{1}{2}$ feet long with one handle, which is comfortable to grasp, and went out to the four splendid live oaks which were killed in the storm, whether by lightning or otherwise I don't know, but they have stood there in melancholy naked grandeur ever since, till this winter I bought a fine cross-cut saw, and had Jim and Joe Keit to saw them down. It was long and laborious, but they had become a menace to the cattle, as the limbs rotted and fell. I selected a limb of suitable height for me to work on and began very awkwardly to saw. The cattle seeing so unusual a sight gathered round me, and Equinox, the bull, feeling sure I must be fixing food for them, came nearer and nearer in his investigations, so that I was forced to an ignominious retreat, before I had made much progress on my "iron" limb. I was not going to give it up, however. I went into the next lot where there was an even more indestructible oak tree, which various men at various times had refused to tackle, and began afresh with the saw. I was pleased to find myself already a little handier and worked with great satisfaction. I remembered Dickens' "'tis dogged does it" and my spirits rose as I got the knack of drawing back the big saw. Jim, who was engaged in cutting limbs from a green live oak, which is much less tough, and which I disapprove of entirely, some distance off, came and expressed great anxiety lest I overexert myself and said, "Let me finish it, Miss Patience, you'll be here till dark," but I proudly declined, and to his and my amazement I had the back log off in half an hour.

"Now," I said, "if I who have never handled a saw before in my life, can cut that log, seven inches in diameter, which has been here since the storm of '93, and rings like metal when you strike it, in half an hour, you and Joe Keit should be able to cut those logs of the same size from those oaks which are rotting a little, in ten minutes, and by giving a day to it, the house will be supplied with back logs for two months at least."

January 10.

I ventured to church in spite of rain which did not amount to much. A little stiff and painful from my prowess with saw yesterday, but would not for worlds acknowledge it to any one. Had my iron log brought in and set up in the piazza, and shall put a geranium on it as a pedestal. I am so proud of it I cannot think of burning it up.

I hear that Gibbie has moved off of the place, has left without paying his rent. He came on the 16th, and paid one dollar on his rent for October, the rent being $1.25 a month, and he says he gets one dollar a day for his work. He assured me that it was impossible for him to pay more until some mythical time when he would be paid off and pay the rest to date. Now he has slipped away without paying at all. I have written to see if I can get it from his employer. Now comes his brother David to tell me he is going — he pays me $2.25 which leaves $6.50 still due me. I made him give me his note payable by April first for that. I do not doubt April first was a most suitable date and that it will be a proper celebration of the day as April fool. Chloe's indignation is great for she knows how often I have

Chloe began: "W'en I bin a small gal."

helped them in sickness and how patiently I have waited on them. She burst out, "De good yu do, de t'enks yu git, how yu help dem po' mudder tru' she long sickness an' tribulation! but w'at better kin you 'spec f'um run-way nigger fam'bly?"

I seemed surprised at this and said, "What do you mean, Chloe?" "Miss Pashuns, you don' know dem kum frum run-way nigger fam'bly?"

"No," I answered, "I never heard of such a thing."

"Well, den, I'll tell you. W'en I bin a small gal, bin a min' chill'un ne* street, my grandpa Moses bin one o' ole Maussa fo'man — him had one gang o' twenty man, en Daddy Sam, Bonapaa't pa, had de oder gang, en dem uster bery proud o' dem gang, en dem gang used to run race fo' wuk. Well, my grandpa had Gibbie grandpa een him gang — 'e name was Able, but Able neber love wuk — soon as de springtime cum en dem biggin fo' staat for plant crap, Able n'used to run way; ebery year de same t'ing — en dat used to mek de gang mad, kase dem had for du him wuk. You onderstan', Miss Patience, dem had to share him task between dem, fu' extry. One time Able bin gone six mont' — de 'hole summa' en Maussa bin a fret, say somet'ing mus' be happen to Able, 'kase him always did cum home befo' col' wedder en now de wuk all dun, en de tetta dun dig een, en we de fix fu' winta! One day de chill'un bin a play in de street en Able gal com on contac' wid anoda' gal, en dem bigin fu' sass one nudda. De oda gal say, 'I'se betta'n yu any way. I'se got Pa, en yu ain' got no Pa.' Den Able gal mek answa, 'I *is* got Pa.' De oda gal say, 'How cum nobody see yu Pa? No, yu ain't got no Pa.' 'I is got Pa, I tell yu.' 'W'ey yu Pa? ef yu got um.' 'My Pa dey up loft een a barrel.' De oder gal tell him pa dat night, en him gone straight en tell my grandpa, en de nex' mo'nin dem tell Mr. Flowers en him tek my grandpa Moses

* "ne" is a contraction of "in the."

en gone to Able house, en dem gone up een de loft, en dey tru'es you born, was Able cumfutable een a big rice barrel! You know dem was big barrel dat time fu' hol' six hund'ed pound. W'en dem tek Able to Maussa him say, 'Well boy, I'm sorry you kyant mek up yu min' to wuk for me, you'se de only run-way I'se eber had, an' if you don' want to stay en wuk fu' me, I'll hav' to sell you I suppose.' Now, Miss Pashuns, yu see Gibbie cum f'um run-way stock, en all o' dem is triflin' no-count people."

Poor Gibbie, I didn't know his ancestral weaknesses, but I recognize the type — quitters all — start with a flourish, but soon leave the track. His mother came of better stock, she was a faithful worker; it is the father, whose name I always spell "Pshaw" because it describes him, who transmits the blood Chloe so scorns. I always have had a weak spot for Gibbie, and now I am more than ever conscious of it. Who of us rises above his inherited weaknesses? Not all, certainly.

Monday.

After all the agitation of Gibbie's disappearance by night, he has returned and entreated me to forgive him, and greatly to Chloe's disgust I have done so and he is back in the stable and I am thankful to have him there.

Tuesday.

As it is impossible for me to stand driving Marietta on foot, I had Gibbie lead her, sitting behind the buckboard, in which I drove her mother. It is absolutely important that she should go out on the public road every day and get accustomed to the sights — to-day I tried the experiment. She went well until I took up the whip, and then she drew back and Gibbie had to get off. I drove on slowly, and fortunately it happened, for just before I reached the bridge I met a white-covered wagon — those country wagons, which, seen so often in the mountains, are rare here, and Ruth was very much frightened by it and would not pass.

If a young man who was sitting by the driver had not got out and led her past I do not know what would have happened. I drove on over the bridge and then back to find what had become of Gibbie and Marietta. I found them still fighting, but after a little patting and talking to, Marietta allowed him to sit on the buckboard and lead her. I went about eight miles, and I hope after this I will have no trouble.

January 11.

Drove Ruth again with Gibbie sitting on back of buckboard leading Marietta. She fought a little about turning down the road, but went ten miles after that at a good rapid pace and gave no trouble, so that I was greatly surprised this evening when Gibbie asked for a few words and said: "I do' wan' to hab' no'tin' mo' fur do wid de colt. I weary wid 'um, en I do' wan' you for call me no mo'. I discouridge 'bout 'um."

I laughed at him about it, but I found he was in earnest and that there was something I did not understand. I said, "You know the colt does not like Dab, and she likes and knows you. When I got Jim to handle her for the month he was here, I would have liked him to take Dab with him to drive Marietta, but he said Dab was not quick enough, and she did not know him — she knows you because you feed her. Now, are you willing for me to go out with only Dab to help me with her?" He only mumbled something about being "discouridge," and I let him go.

January 12.

It is a perfect spring day; it is hard to believe we have two months of winter yet. Of course I could not give up taking Marietta out because of Gibbie's whim, so I ordered the buckboard with Ruth, and Marietta with halter to lead behind as usual, and seeing by his stolid, sulky expression that

Gibbie had not changed his mind I called Dab to lead Marietta. We got off better than I had expected, Gibbie looking on with a Mephisto expression. Things went very well until we had gone about half a mile, when an old mammy with a shining tin bucket in her hand came out of a side road. She made me a deep curtsy and went on, I supposed, when I heard her exclaim, "My lawd, 'e git 'way," and looking back I saw Marietta flying down the road, with the long halter twisting about and Dab in hot pursuit. What was I to do? Ruth will not stand. I got out and took the halter and laboriously sought a tree which would suit by the roadside and tied her. Then I flew down the road, calling to Dab to come back and not pursue the colt. At last he heard, and I sent him to stand by Ruth and I walked rapidly after Marietta. She was out of sight, but at last I came to the place where she was grazing by the road. When she first saw me she moved off, but I stood still and called her to me with many blandishments and promises, and she came quietly up to me, let me take the halter and lead her back to where the buckboard waited.

I charged Dab not to let it happen again, and we drove on without further adventure. I told Dab he need not mention his great carelessness to any one; that I was too much ashamed of it to wish it known. I hope Gibbie will never know, as we came home all serene, having been to Miss Penelope's and made many necessary purchases.

January 13.

When I went to the stable this morning Gibbie had already taken Marietta and led her down the road with the blind bridle on! I was greatly surprised and amused. He had thought to scare me, thinking I would not be willing to take the colt out without him, but having failed in that he has returned to his allegiance to her. Jim put a bridle on her without blinkers, and it has made her very difficult to manage. I have not been able to use a whip at all. I cannot lift my hand to my head without her jumping, so that I am perfectly delighted that Gibbie put the other bridle on her. I do hope I can soon have the harness, which has gone to be mended, so that I can drive her again.

January 18.

Went to Casa Bianca, where things are in a bad way — the hands positively refuse to come out to work when called by Nat. There is no one I can think of whom I could make foreman. Nat works faithfully himself and keeps his accounts straight, and if the hands will not accept him they will have to go. From the time Marcus left they have done nothing. They planted five acres of rice-land apiece, but did not work it at all, so that they did not pay their rent, and I know they would do worse this year. It has proved a splendid crop year, and they could get $1.15 a bushel for their rice, but they have none, because they were too lazy to work it. They grumbled and jawed about "not takkin' orders from de young man I put in charge," and when I asked point-blank if they refused to take orders from my foreman they answered that they did, and I told them to leave.

These men will go to my neighbors, who will be glad to have them, and I trust they will improve and get back to the point they had reached when Marcus left. It seems a pity to have such beautiful lands as I have there, lie idle for want of hands.

I told Nat to do the best he could with the few left and to exact a shad a week from the fishermen who are now spreading their nets in the river just in front of the house.

January 23.

Got into road cart at the front door and drove Marietta down the avenue for the first time. She went well; it was very hot and she was in a great heat when we got home. Went down the road to the log school-house, and no wellbroken horse could have done better. With joy I gave her her two quarts of oats mixed with hot water and soda; this has nearly cured the lampas from which she was suffering. Waited dinner till 4:30 o'clock, expecting Mr. G., but he did not come.

January 25.

The hands all pulling corn-stalks; Gibbie hauling manure to corn-fields. I did not stop him to drive Marietta until 1 o'clock. She behaved very badly at the turn of the road going to Peaceville; she wheeled suddenly and reared at nothing that I could see. Gibbie held on to his seat. He said, "'E smell goat, I smell um meself." I think she was provoked at not getting out sooner — we generally go just after breakfast. Went on to Peaceville and made a visit, but she was very ticklish all the time and on the way home she tried to run twice. As we got nearer home she quieted down, and I knew she was thinking of the oats, but I did not give it to her, for she understands perfectly. To-night I finished the first volume of the "Life of George Eliot," by J. W. Cross, which Mr. G. lent me. My sympathy with her is great. A grand woman in mind and heart. Such a misfortune she should have fallen under the Bray influence.

January 26.

A most exquisite sky at 6 A.M. and a wonderful sunrise. Thank God for all His beauty!

Drove Marietta down and took lunch at Mrs. H.'s. She went beautifully. Stood quietly the hour I was there, scarcely moving, and was as gay as possible at the end of the sixteen-mile drive, and I gave her her reward with delight.

February 2.

Have had the pleasure of a friend staying with me, and my diary is blank in consequence. While my friend was here I could not drive Marietta and very much feared that the week's idleness would make her unwilling to go quietly this morning, but she did remarkably well. We just escaped terrible danger in the shape of a party of boys driving a team of goats. I saw them in the distance and was wondering what I should do, when they turned off into another road. Coming home for the first time she had to come behind a buggy, which passed me while I was stopping at the post-office. She did not mind it at all. It was a great satisfaction to me, as the occupant of the buggy was one of my dear neighbors who had predicted terrible things if I undertook to break the colt, and had said, "My dear Mrs. Pennington, at your age you ought to have more sense than to do such a foolish thing."

CHEROKEE, February 18.

Drove Marietta this morning and she behaved like a fiend. With all my heart I thank the good Father for his great mercy to me.

She started off pretty well, though I felt a subtle something unusual about her.

In a woman it would be called "nerves." About a mile up the road she had settled into the long, swinging trot, when through the pine woods running toward us I saw two little darkies in startlingly red frocks and startlingly white pinafores. This only was needed to upset her. She jumped, she pitched, she went from side to side of the road, but she did not get away from me, and after a little fight she quieted down, and I called the two little girls, who stood dismayed near the road, to me and talked to them, as that is always the most quieting thing to her. She seems to listen eagerly, as if trying to understand. After a few seconds we went on, very gingerly at first, but soon she resumed her beautiful level trot, head up, nostrils distended, and speed gradually increasing as we went on. It was delightful, and with a sigh of relief I shook dull care from me and gave myself up to the enjoyment of the moment — the perfect day, the battle won, and the beautiful, sleek bay creature, whose every pulse and thought I seemed to feel. Suddenly I saw fifty feet ahead at the opening of the Hasty Point avenue, where the grass stood high, two black heads rise above the brown, waving sedge a second and as suddenly disappear. Just as I saw them Marietta did. She stopped short, almost throwing me on to her back. Then, quick as lightning, wheeled and bolted, putting the left wheel into the deep ditch, throwing me so far out on that side, that my ear felt the wind of the wheel, and was spattered with mud though not cut. Luckily I had a firm grip on the rein with my right hand, and having learned to ride by balance, I did not go out, and my whole weight going on that rein, pulled the left wheel out of the ditch as she ran, but it was a near thing, and God's great mercy. She ran half a mile before I could pull her down, then I turned and drove her back, finding Gibbie on the way. He was thrown off when she wheeled, and of course could never catch up. I drove her ten miles and then up the avenue, where she had been frightened. The two boys (16 and 18), who had caused the trouble, came up to me and begged my pardon. I spoke severely to them, for some years ago I remember they scared the mail man's horse in the same way, and so could not plead ignorance. He, being a man of action, shot at them, frightening them terribly, and yet they have done exactly the same thing again, though a man who passed them in a buggy, warned them that I was coming with the colt.

February 20.

It rained yesterday, so that I could not drive Marietta, as I wanted to do, but I took her out immediately after breakfast this morning. I could not go as far as I wanted because about four miles down the road I found all the woods on fire, so I turned before she got too frightened. She was very good, so I gave her both oats and potatoes when we got home. To-morrow is our rector's Sunday and he is to stay with me. I will have to use both buckboards. I had Bonaparte put a new seat to the old one, and it looked so badly that I could not resist painting it when I got back from driving Marietta, thinking I would have time if I worked rapidly to get through before Mr. G. arrived. I was so absorbed that I did not hear the noise of the rowboat coming, and so he found me in my big apron hard at work. I was sorry to be caught, but the job was finished and looked very fine — at least to my eyes.

Sunday, February 21.

A very pleasant service. Mr. G. was to go on to the mission service for the pineland people in the woods about nine miles away, so he lunched in Peaceville and I returned home. As soon as I got in the gate Chloe called out "Good news," and I found to my delight that A. had run down from his legislative duties to make me a little visit. Such a pleasure!

February 23.

Got up at 3:30 to have coffee and toast for A. to go out ducking. If you do not go early there is no use to go ducking at all. We had lunch at twelve, and then I drove him to Gregory to take the afternoon train. He got twelve English ducks, which looked very imposing as he got on the train.

February 26.

One of the road cart wheels is dangerous, so I had Romola put in and took my side-saddle along, and drove up to a man about six miles away and left the cart to be mended. Dab swung on the little shelf behind and saddled Romola for me, then walked home while I rode around by Peaceville for the mail. It was a long fatiguing day, but beautiful, the only drawback being that Don, my splendid red setter, came upon a swarm of little pigs about three days old and killed one of the tiny things, and the old woman to whom they belonged was much distressed. I gave her what I had in my purse, but it was not much, and that pig meant such immense hopes! I felt for her — oh, the pitiful little realities on which we build such towering hopes!

March 1.

I took the whole household down to Casa Bianca to-day — Chloe, Patty, and Dab — for I was giving a luncheon. It was a charming day and the place looked fascinating to me, and every one said the same thing. I took out and used all my beautiful china, which I rarely do, because it is such a critical business to get it all washed up and put away before leaving. That is why I took Chloe; that, and the hope of getting a shad fresh from the river and having it planked. One of the guests was from the North and I wanted her to taste it fresh from the water, but alas! Nat was so occupied getting himself dressed in a stiffly starched shirt and other unusual adornments that he did not get the shad. I was greatly disappointed, but we went after lunch down to the river in front of the house and saw some caught, and we each carried one home with us.

March 3.

Have not been able to drive Marietta, because the road cart had gone to be mended, but to-day sent Dab up to get it. Have had a great deal to worry me. I had a letter from the matron to tell me that my poor, dear little darky Rab is ill of typhoid pneumonia. She says he calls for me all the time, and asked her every day if she had written, so she had to write. Last week I sold a steer to a man for $13. He declared he had the money or I would not have sold the steer. To-day he arrived, bringing $4, with voluminous promises of the rest in a month's time — it will probably

I took Chloe to Casa Bianca to serve luncheon.

be six months before I see the rest of the money if ever, and now I want to send money for Rab's illness.

March 5.

I sewed until 11:30 and then Gibbie brought the colt. It was a perfect day and a joy to be going out with Marietta again. She threatened trouble at the gate, but Gibbie ran to her head and I gave her one or two sharp cuts with the whip and she went on, rather sulkily however, so I had Gibbie walk ahead as far as the bridge, but did not see until I was just up to the bridge a huge flat with a house on it, a great smoke coming out of a pipe on top, half under the bridge. I called to the man angrily to come out and speak. Marietta seemed squatting with a view to some desperate action, and there was nothing I could do. I could not force her over the bridge in a state of fright — it would have been most unwise — it would have been equally unwise to turn around even if I could have done it, so I appeared to have forgotten her and told the man it was against the law to tie his flat under the bridge in that way, that it was enough to frighten any horse, and that was actionable, and besides that it was very bad for the bridge, which was a great expense to the county, so that he could be indicted on two counts. I was delighted with my fluency and at its effect on the man, but kept my eye on Marietta, who was on the point of wheeling but was too much interested to carry out her intention. The darky was most apologetic and polite and explained that it was not by his desire he was a fixture under the bridge, but that he had stuck there as he tried to get through. "Worse and worse," I said, "as the tide rises you will carry off the bridge entirely!"

He did not know that I was talking really for the galleries, which meant the colt, though I felt provoked with the man for trying to get under the bridge with a flat too broad and a two-story house, you might say, on top — my own bridge over the same creek had been carried off in that way two or three years ago, and I found it would cost $200, too much, to have it put back, so that my sheep and cattle are entirely cut off from 300 acres of woods pasture, and that is a great loss to me; still I know too well the futility of words under such circumstances and it was merely to make time for Marietta to take in the unusual sight — the man explained that he hoped the tide would not rise any more and that when it began to fall he would try to back out and not undertake to get through again. I asked if there was any one else in the flat. He said yes. "Tell them to come out and let me see them and hear their names." So a man and a boy came out and stood on top of the house in the smoke and I lectured them as to the great blessing of having sense and using it. By this time Marietta was so deeply interested that she relaxed entirely and at once I shook the reins and told her to go over, which she did as quietly as possible, which I think was wonderful. I don't believe any power would have taken either of the other horses over, but Marietta is so reasonable. I took a long drive and by the time I came back the flat had gone.

March 7.

Had another letter from matron of Jenkin's establishment saying Rab is better but very weak and always calling for "Miss Pashun," and I have made up my mind as soon as I think he is strong enough to go down and bring him home. I have for a long time been suffering from a tooth, but felt it a great extravagance to make a trip simply to go to the dentist, but now that I must bring Rab home I will combine the two. I have been very worried about money and very miserable.

March 9.

Took the party down to Casa Bianca for the day, which we all enjoyed. Just as we got to the gate in the very narrow lane bordered with rose-bushes on each side, which takes the place of avenue at Casa Bianca, met a very large white-covered country wagon with four horses driven by two white men. The horses were terrified and it was very hard to get by without breaking up things.

I asked the men what they were doing in there, as it was strictly private property, and not on the way anywhere. The men were surly and refused to answer, and when I asked what they had in the wagon they still refused to answer and were disposed to be rude.

I have been much tried by having the plants in my once beautiful garden carried away and I feared this was a depredation. I stepped out of the buckboard, in which I was driving behind the wagon with the wise men, and walking to the great covered wagon parted the flaps and looked in. This seemed to enrage the man who appeared the owner, and we had quite a scene.

I was completely satisfied by my inspection, and when I explained my motive for wishing to know the contents the huge, florid wagoner seemed quite ashamed of himself for not having given a civil answer to my question as to what had taken them so far off the public highway and into my private grounds. He now vouchsafed the answer that he had gone in to buy a shad, and with many apologies on his part and much admiration of the beauty of the place we parted.

My wise men were most enthusiastic over the garden, where the camellias were in full bloom, though the azaleas were not yet out — Mr. Poinsett planted this garden somewhere between 1830 and 1835, was a scientific gardener and brought many rare plants from Mexico, among others the gorgeous Flor de la Noche Buena, which has borne in this country the name Poinsettia in his honor. There is very little left of the original garden, only the camellia bushes which have grown into trees and the Olia fragrans, Magnolia purpuria, and Pyrus Japonica. The cloth of gold, Lamarque and other roses which grew rampantly, rejoicing in congenial soil, have been carried off from time to time by visitors, and the hedge of azaleas has been almost destroyed in the same way.

Nat is watchman there, but of course he cannot prevent such things; he can only remonstrate. Thus far he has been able to protect the house.

I was quite touched by the interest of Mr. S. He was

much impressed by the books, prints, etc., which have been shut for thirty years in the house and of course moth and rust have corrupted and done their work. As he looked over them he got quite excited and said : —

"Mrs. Pennington, say the word and I will send to Gregory and get boxes — this young man can go at once for them — and I will pack all these things for you and ship them to the north and sell them for you to the best advantage."

When I demurred he added : "It shan't cost you a cent; it will be a pleasure to me to attend to it. These things interest me and I cannot bear to see them perish. They are valuable and could bring you in a good sum."

I said: "I am much touched at your kindness, Mr. S., and thank you very much for your offer. I think you greatly exaggerate the value of these things. I sent on some of the most valuable this year to New York and got a pitiable result in money. I knew those things to have a value of about $600 at the least and I got $100.

"You would take all the trouble and expense of packing and transporting these things and when you went to dispose of them you would find nobody wanted to give anything for them, and the greater part would be treated as rubbish. You would be embarrassed by your effort to do a kindness.

"No, let them stay where they are, where they have a right to be, and where if they are rubbish, they are at least rubbish dear to my heart."

It was hard to make him accept a refusal of his kind offer. It is not the first time that offer has been made to me. I greatly appreciate the kindness which prompts such active interest, but I cannot accept it.

I cannot place my dear old possessions in such a position. Let them grow old comfortably unexposed to comment and criticism and above all appraisement. I do not defend my position — it is unreasoning and I suppose unreasonable;

but unfortunately I am made that way. Mr. S., who is a practical and very kindly soul, was quite distressed at his failure to convince me. The commercial instinct is lacking in me altogether, I fear.

CHEROKEE, Sunday.

It was a great effort to go to church this morning, but I went and was rewarded. I enjoyed the service, and the short sermon was beautiful, on the 22d chapter of Genesis — Abraham's call to sacrifice his son, his only son, Isaac. It seems Isaac means laughter. Abraham in his great joy at this unexpected and belated blessing called the child Laughter. That makes the story more wonderful.

GREGORY, Monday.

By to-day's mail I got a letter to say that Rab had been sitting up a week and for two days had been out of the house, so I suddenly made up my mind, ordered the buckboard, and told Bonaparte to prepare to go with me and drove down.

I had two delays on the road, one of about ten minutes at the ferry, and as I had left very late I missed the train by three minutes. I had driven six miles in a pouring rain. As it was bright when I left, and my buggy umbrella is faded and torn I had left, that good friend at home, so I had to take the rain protected only by a small umbrella. And then to come to a hotel where there was no fire or means of warming!

However, these are occasions for showing one's philosophy, and I have not fretted at all, but amused myself imagining what it would be to live in a hotel, or hostelry of any sort, permanently. The thought made my strenuous, and sometimes a little hard, life seem ideal in spite of its limitations.

CHEROKEE, March 19.

Returned from Carrollton last night and was most pleasantly entertained at Woodstock. Brought poor, thin, shaky little Rab up as far as Gregory, having written to Jim to meet him at the train and take him to his house for the night.

The child seemed overjoyed to be coming home. Dab brought the buckboard and pair to Woodstock, without any catastrophe, at 10 o'clock. I drove into Gregory to Jim's house to pick up Rab.

I found him still beaming in a very feeble black way, and still grasping the coverless shoe box with which he had appeared at the station. Jim's wife said she had been glad to have him spend the night there, and her mother, who belonged to one of our most trusted families in the far past, came out and gave me a very beautiful blessing, which went to my heart, and at the same time made me laugh, as she began: —

"Po' little man! will ondertak t'ing too big fur um! But de Lawd'll bless um all de same," and so on indefinitely.

When we reached Cherokee the mystery of the shoe box was revealed. With trembling fingers Rab unrolled a gorgeous cup and saucer, rose adorned and with a heavy gilt band, which he presented most awkwardly to Chloe, and after some fumbling in the newspapers of the box produced a mint candy basket filled with broken bits of candy which he poked into Dab's hands. The effect was dramatic.

Chloe had not pretended to be glad of Rab's return, and her greeting had been cool, to say the least. Now she was so surprised as to be quite overcome. Dab had said to his confidants that as soon as I brought Rab home he would leave, for he knew he could not keep good with Rab here. The candy had a most pleasing effect upon him, so poor little Rab had a cordial home-coming at last.

When I went to the orphanage to see him and the arrangements were made for him to meet me the next day at the train his look of tremulous joy at the prospect of going

x

"home" was very pathetic to me, for I knew I was the only creature who would greet him with pleasure there. I took out a quarter from my very empty purse and said: —

"Wouldn't you like to buy a present to take to Chloe and Dab?"

He answered with delight that he would. When he met me at the station with the very respectable and pleasant matron, his well-worn valise beside him very much stuffed out and the shoe box covered with newspaper tightly held in his hand, I supposed that was his lunch, but at Lane's when I asked him if he had any lunch he answered no, and I gave him some of mine. I wondered over the contents of the very unhandy package, but did not inquire about it.

I was greatly pleased at the success of his offerings and I think he chose very well.

CHEROKEE, March 20.

Wrote furiously for the mail, and by the time it came at 11 had ready letters containing checks to pay off all my debts, which is an immense comfort, though accomplished by the sale of things very dear to me. I am thankful now that is over.

I wanted to drive the colt, but felt too weak and worthless, not to say confused and discouraged, to attempt it.

March 21.

Drove Marietta to Peaceville and then in to Miss Penelope's, making about ten miles. She wanted to fight twice, but when I spoke to her and said "Mind your oats" she steadied herself and went beautifully, and I had the pleasure of giving her a generous portion of oats.

Gibbie gets out fifteen ears of corn for her at a feed, but I fear me she never gets more than five. The product of the patch behind his house planted in corn is unlimited. He is still selling corn weekly, ostensibly from it.

Bought fishing tackle, lines and hooks from Miss Penelope this morning and hired old Tiny to come and fix up the lines and go out fishing with Rab, hoping for a fish now and then to eat, and that it would prove a most peaceful, healthful way for Rab to pass his time, until he gets stronger.

March 22.

Went out to street to visit Gibbie and see if he was really ill or not. Found him sitting by the fire. I don't know whether there is anything the matter or not. Went to see Elihu's little daughter Juno, who is in a very bad way, so weak and emaciated that it is painful to see her.

March 23.

Ransom came to-day for money due him for making the chimney for the house I had fixed for poor Elihu to move the remnant of his family back home. I can ill afford it, but I thought the march of death might be impeded by their coming back where they were born, and besides I can help them by sending things to the ailing ones.

Ransom talked a great deal. I sympathized with him in the death of his grandchildren, Estelle's children. They seem to have developed a new disease which has puzzled the doctors — some acute condition of the eyes, inflammation producing blindness and eventually death. We have had for some years a clever graduate of Johns Hopkins in this region who is making a study of malarial diseases. He went North three months ago, and one of the negroes telling me of an illness when they had to do without the doctor, there being none within fourteen miles, said with an air of intimate understanding : —

"We doctor gone fu' larn fu' scrape eye. 'E say him don' kno' nuff 'bout dat, say him neber larn fu' scrape eye yet."

Ransom talked on, giving me the news of the colored world

and the crops, etc. It consumes much time, but I try to lend a willing ear. Finally he said: —

"Miss Pashuns, I got a great tenks to gi'e you. You don' me a great good. Maybe you don' fu'git, but I 'member. You kno' dat time I bin een sitch big distruss? I los' me wife, I los' me ox, I los' me cow, en I come to you fu' help, en you mek answer en say: 'Ransom,' says you, 'I ain't got no money to gi'e you, but I kin p'int you to help. Wot's happen to you is happen befo' to anoder puson. Now you go home en tek yo' Bible down en look fu' de book o' Job, en you mek a prayer to de Almighty to open yo' mind fu' onderstan', en you read de book o' Job en study ober him.'

"Dat was yo' discose to me, en I gon right home en I tek down me Bible, en I fin' de Book o' Job; en, Miss Pashuns, I was dat 'stonish! Dey was all me feelin's, en all me sufferin's, en eben all me wud, rite dey; en I read, en I read tell de kumfut kum to me. En, Miss Pashuns, ma'am, my min' bekum quiet en happy en I neber is fret sence. So dat wus a presunt yu mek me dat time abuv gol', kase 'e kyant loss."

I was greatly amazed and touched, and I said: —

"Well, Ransom, you have returned the gift to me, and I thank you, for I have been terribly worried and harassed in mind and spirit, and you have brought to my mind where I can find help. I will turn to the Book of Job myself to-day."

Having begun on a real discourse, Ransom was not willing to stop. He went on: —

"Anoder t'ing I wants to tell you, Miss Pashuns. Las' Sunday week five o' we mens, all mauss nigger [negroes once owned by the same person; it is a bond of fellowship], meet in de road, en Joseph say: —

"'I wants to tell unna ob a wision I had. Las' nite I wake wid a big light een de rum, en I rub me eye en I look, en dey I 'see ole Miss; 'e stan' en 'e look on me — 'e look nyung, 'mos' like a gal, but you cud tell rite off 'twas ole Miss, kase 'e had de full look o' she een 'e eye, en 'e dress was all w'ite en shine same like lightnin'; 'e wus too butiful. I look en I was dumb; 'e neber say not'ing, 'e jes' look at me so kynd en den 'e fade 'way. Now I wan' to kno' wha' dat signify. 'Tis a tokin fu' sartain, but wha' does 'e signify?'

"En I mek answer een dese wud: 'My bruder, 'e is a tokin f'r good sho'ly. Ole Miss is een Heben es sho' es you bawn.' En 'e say, 'Yo t'ink so? Yo' t'ink ole Miss is een Heben?'

"En I mek answer en says, 'Ef ole Miss ain't een Heben, den no mortal man or 'oman ain't dere. Now, Joseph, you kyas yo' mind back, en recomember how ole Miss fight wid we all fu' teach we, f'um de time him married ole Maussa — en dem was nyung den, en 'twas my pa dem bin teach den — ebry libing Sunday ole Miss hab ebery chile on de whole plantation en teech dem. Fust 'e teech "Our Fader praise," den de Ten Kummanment, den de "I belieb" praise, den w'en we kno' all dat, sose we kin say um widout stop, den 'e teech de wud o' de blessed Sabior, chapter at a time, till all we chillum w'at cudn't read, we hab we head chock full o' Scriptur.

"'Now w'en we dun say we Katakism den up kum Maum Mary wid de big cake een de wheelbarrer, en ole Miss kut um 'eself, en gib eech chile a big slice. I neber tas' sech cake sence, 'e had su much aig, en su much sugar, en su much short'nin' 'e mek me mout' water now, w'en I t'ink pun um.

"I read tell de kumfut kum to me."

"'Now, Joseph, I ax you if ole Miss ent mek she title clear to him manshun een de sky? 'E cud a bin a sleep, or 'e cud a bin a dribe out een de open karrige fu', wisit she fren', or 'e cud a bin a eat cake sheself, but no, Sunday afta' Sunday, kump'ny or no kump'ny, fo' o'clock Sunday ebning yu'le fin' ole Miss een de church Maussa build een de abenue, wid f'um fifty to one hund'rd chillum de wrastle wid dem ondirstand'in'.

"'I kin read now, but my breder, all de fulness o' my min' kum f'um dem Bible wud dat I got. I don't need no spectacle, I don't need no light, I kin jes' pore out de Scriptur to eny po' sinna I meets nedin' um.'"

I cannot give any idea of the balm these simple words brought to my bruised and wounded spirit. I thanked Ransom with all my heart for his beautiful, earnest testimony to my dear mother's unwavering devotion to her duty as she saw it, from the time she came to the plantation as a bride of nineteen.

Before going Ransom wished me many blessings, and wound up by saying, "Miss Pashuns, I hope you is conwert?"

Quite alarmed, I asked him what he meant.

"I mean I hope you's got religion, ma'am."

"Oh, Ransom, I hope so."

"Well, ma'am, I'm glad to hear it, en I hope 'tis true."

He did not seem to feel quite satisfied about it, which was a great shock. I know, measured by the standard he had in mind, I fall very short. I must fly to Job at once.

CHEROKEE, March 24.

I have had the great pleasure of a short visit from my friend M. T. She had only a few days of rest from her work in the East Side Settlement House, and to my refreshment and delight she came to me. I love to hear of all the wonderful work done there.

"Up kum Maum Mary wid de big cake een de wheelbarrer."

On Thursday I had a most surprising letter from an unknown friend in New York, saying she had become interested in the children of my Sunday-schools and asking if she might send some little Easter presents for them. It was so unexpected and so delightful! I had no thought of being able to get anything for the children.

I wrote her at once, giving a list of the children of the three distinct classes in which I am interested. There is the class of little gentlefolk in the hamlet of Peaceville whom I teach in summer first, then the larger class at St. Peter's Mission Church out in the pine woods. These are the children of the white workers in turpentine. Finally there are the little darkies on the plantation whom I teach in winter, when I can get them. Their own churches, Methodist and Baptist, are very jealous and discourage their coming.

I wrote Miss W. that I sent them all, so that she could choose the class to which she would send presents, and told her how to address the package.

It rained heavily in the afternoon. Gibbie did not come, so I had to milk. I was perfectly delighted, because I got more milk from Winnie than either Gibbie or Dab has been getting. When I was in the mountains one summer I took regular lessons in milking, for the mountain folk milk beautifully, whereas the negroes are generally poor milkers. They never can take all the milk, and if you do not keep the calf to take the balance when the milking is over, the cow will go dry in a very short time. Leave a pint to-day and to-morrow there is that much less, and so on, a pint less every day. The cow is soon only fit to turn out to pasture.

You cannot teach what you do not yourself understand, so I took milking lessons, and as a teacher have been rather a success, but have been generally greatly mortified at the results of my efforts at milking myself. Hence my pride when Chloe said the milk was much more than usual. Chloe cannot milk, she draws the line there, and Bonaparte is still working on the pineland house four miles away, and does not come to the yard at all.

After taking the milk to the house I went to the barn-yard and fed the oxen. Gibbie had taken them out of the plough and turned them out in the rain with nothing to eat and had gone home. I gave them a good supper and then went home, changed my wet clothes, and had my tea and toast and then a delightful evening reading "The Power of Silence." A wonderful book, to my mind.

Gibbie and the oxen.

March 25.

Good Friday. B. and her dear little party arrived safely at 1 o'clock. It had poured all night and part of the morning, so I was anxious about them. The children are lovely, the baby like a sweet flower with her heaven touched blue eyes. Unfortunately their trunks went astray in some way and Dab returned with the wagon empty, except for the baby carriage.

Easter Sunday, March 27.

A beautiful day and charming service. The collection was for missions and our delight was great at finding it was a little over $12. It will pay up our apportionment. I drove our rector to church in Peaceville and then let him have the buckboard and Ruth to go on to St. Peter's, while I came home with my dear little neighbor, who dined with me.

A number of little darkies came to Sunday-school and sang very nicely. Lizette came for the first time. She is about 14, very tall and gawky, but with a good face. She knows not a word of the catechism, while Goliah and the other little ones say their Creed, the Ten Commandments, and that most comprehensive duty to God and duty to one's neighbor, glibly. The Easter hymn which they have been learning for a month, "Christ the Lord is Risen To-day," went beautifully. They left with great speed after receiving a double portion of candy in honor of Easter.

After they had gone, I went out to enjoy the exquisite afternoon, with its rosy golden light, and there at the foot of the steps was a huge snake. I looked for a long pole and killed it after a fight. While I was finding the stick it had got under the house, which made it harder to kill it. It did not seem quite dead and the puppy wanted to play with it, so I went into the yard and got the axe and chopped off its head, and as Prince, who has no country sense, still wanted to get the head, I buried it quite deep, all of which somewhat interfered with my enjoyment of the peace and beauty of the Easter gloaming. It makes one think, when these terrible discords come into the harmonies of a perfect day, must the trail of the serpent creep into everything?

Am I yielding to the temptation of getting too much amusement out of my dusky little scholars? Do I not agonize over them sufficiently? That may well be. It seems so hopeless to reach below the surface, so hard to influence the spirit, the life, by this hour's teaching once a week. Still I must do what I can; I cannot see them follow their blind leaders without making an effort to help them.

It does not come to them as it does to the heathen, who have never heard of God, as something new, a revelation. They hear great professions of religion and calls upon the Lord, and yet there is the daily example of deceit, faithless work, the snatching up of any and everything that can be stolen unseen. To be discovered is the only sin; you may lie, break any of the Commandments, only don't let it be found out. This going on daily, hourly, yearly, who but the Holy Spirit can contend against it? I constantly tell them that when they have all these good words stored in their minds, or rather their memories, and at their command, they have only to call them up, when Satan attacks them, to be able to defeat him. Even our blessed Saviour when tempted by the Evil One did not answer in his own words, but in the words of Scripture. "It is written," was the preface to his reply. I often feel that this is the greatest thing we can do for children, to store their minds with these powerful words, which will come to them in their hour of trial, as weapons against the deadly spiritual foe.

Sunday, April 3.

To-day I had the joy of distributing at St. Peter's-in-the-woods the pretty Easter eggs Miss W. had sent for the children. It was a joy to see the usually phlegmatic faces light up at the sight of the lovely things in the familiar form of an egg.

I asked the very pretty young mother who tries to keep the Sunday-school going all the time, though as she says she "has mighty little knolidge herself," to tell me the name of the best scholar. She answered very demurely: "It wouldn't do for me to tell you, Miss Patience; the best plan is for you to listen to the lesson an' then you can tell yourself."

When the lesson was said I found her little boy of 6 was far ahead of the others in saying his lesson and that was why she could not tell me. The next best was a boy of 14 who was, she said, the most punctual of all in attendance, coming a number of miles on foot in all kinds of weather, but he had no one at home to help him with the lesson.

"So, in reason," she said, "he couldn't know it as good as my little boy, fur I teaches him; but Joe does his best, en he aims to learn."

So I decided to give the rabbit about five inches high to him, and said: "Mrs. M. tells me you are so punctual in coming, Joe, that I am going to give you the largest one of the pretty things, which a kind friend away off in New York, has sent as an Easter offering."

To see the heavy, patient looking face suddenly light up and then fairly beam, when the rabbit was put into his hands, was too delightful. I did not look at him too hard, it was such a revelation of fourteen years of limitation and privation unconsciously borne.

I passed on and gave each child a most beautiful egg. They were all filled with little sugar eggs of different delicious flavorings.

To the children of the cities these things are all well known, but to these little pine wood children of nature they were heaven sent mysteries. When I had finished the distribution the big boy Joe came to me and said: —

"Hear, Mrs. Pashuns, my rabbit rattles!"

"Yes," I said, "he is full of little eggs."

"Will I have to break him to get at them? Fur I'd ruther not get um than to break him."

When I showed him how to take the head off, his content was complete.

Got home just before dark, tired and very hungry after the eighteen-mile drive and the two services, but having thoroughly enjoyed the day. It was very pleasant that it was our rector's day with us, so that I drove him out to the church and back instead of taking the drive alone.

CHEROKEE, April 4.

I am worrying as to how I am going to get seed rice. Some hands want to plant a field of rice, and it seems to me I ought not to be behind them in faith. If they are willing to risk their work, I ought to be willing to risk the seed rice. But the question is where to get it.

The great destruction of rice by the floods last summer

In the field — sowing.

has made seed rice very scarce and very high, and of course no one will be willing to sell it except for cash. However, I have promised to try and get it for the hands who want to plant Vareen.

While we were sitting at breakfast this morning Chloe came to the door and mysteriously beckoned to me. I rose at once and went out knowing something had happened by her tragic expression. When we were out of hearing from the dining room she said: —

"Miss Pashuns, Rab is shot 'eself."

"Good heavens, Chloe! Where is he?"

"Right to de pantry do'."

I flew out and there was Rab moaning piteously with the blood streaming from his left hand. It was no time to ask questions. I called for a basin of hot water and sent to my room for a roll of absorbent cotton and a bottle of turpentine and washed the wound, which was all burned with powder.

The missile, a jagged piece of lead, had gone straight through the hand, making a very ugly, ragged wound. How it got through the muscles, veins, and bones between the second and third finger without touching any of them is a wonder. The bleeding was not excessive. I packed the hole with cotton saturated with turpentine, both top and bottom, getting it as far into the wound as I could.

Poor little Rab behaved very well, did not scream, only the tears rolled down his very black face. After it was bound up securely, my niece fortunately having a roll of bandages with her, I asked him how it happened. He said he was playing with the plantation musket, trying to get out a piece of lead that was in it. He had the palm of his hand over the muzzle when he moved the trigger, with this result.

I did not scold him; what was the use? All my efforts to give him healthy and satisfying amusement and occupation in the boat have been in vain. He will not go with old

Tinny, nor to fish at all unless Dab leaves his work to go with him. He is too weak to do any work, and there is nothing that he can be persuaded to do but play with some firearm.

Dab has done wonderfully well, for the house, of which I am generally the sole occupant, is now quite full, and Dab has the dining room work, which he does beautifully. He has confided to Chloe his disappointment about Rab. I have been terribly disappointed myself, but tried not to write about it, indeed I have tried to ignore it altogether. The child has been ill and got somewhat spoiled, as all sick people do who have any kind of good nursing, and then he is so weak and miserable now. Two days ago Dab rushed into the kitchen in great excitement and said: "An' Chloe, Rab is de very debil self! Not de debil son, nor him brudder, but him very self."

Chloe was delighted to sympathize on so congenial a subject and went on: —

"Rab los' all the manners he carry frum here, an' he ain't brought nutting back."

Poor little Rab during his five weeks' illness has got spoiled, and with his physical weakness, his temper gets the better of him more and more, that is all.

April 5.

A dear little cousin arrived this morning to make a long-deferred visit.

I found Rab's hand looking so ugly and swollen when I went to dress it that I determined to send him to Dr. G. in Gregory, for I am sure it needs a doctor's care. The hospital has been closed for lack of funds to carry it on, but I wrote to the doctor, who I know will do his best for the child. I wrote also to Jim asking him to keep him at his house, and I will pay him.

While I was writing the buckboard was being got and I drove Ruth as hard as possible to catch the mail man. I knew he had left Peaceville by that time and I had to calculate where I could strike him on the road. This I succeeded in doing, and put Rab in his charge, to be taken to the doctor at the hospital where he lives, though it is closed to patients.

It was all very fatiguing and exciting, and my heart was very sore for the poor little piece of black humanity, who has such terrible things to contend against within. I am so glad I was able to send him down at once. It has all broken in somewhat on my enjoyment of my guests, but I hope now the unusual excitements are over.

April 6.

This morning when I came downstairs I was surprised to see the table not prepared for breakfast, as Dab usually has finished all his dining room work by the time I get down. Chloe said she had sent Patty out to knock on the door of his house twice to wake him. She had knocked hard but he would not come out.

I walked out to the house to see if he was ill, opened the door, and he was not there. His valise which he always kept packed was gone, also the fine red blanket, which I bought back for him when he sold it last winter, was gone.

Though I had so often told him when he wanted to go just to tell me, and I would write a paper stating his capabilities and good qualities, so that he could get a good place, he had slipped away in the night! I was quite knocked down by this. The excitement about Rab had taken a good deal out of me, and now I was dismayed.

The house is full, and though Patty is a good little girl and specially eager to wait on table, she knows very little and my whole time during a meal has to be given to seeing that she does not lose her head, and do something very unusual, to say the least, but I now called her and told her she must take charge of the dining room, in addition to her other work.

She showed all her white teeth and expressed delight at

that. I showed her exactly how to lay the breakfast table and what dishes to use, and then went up to my room to compose myself before the family came down.

Gibbie having determined to take a rest, as he very often does, had announced himself sick and Dab had been taking care of the horses as well as the cows. Fortunately Gibbie came out this morning, but when he came for the stable key I found that Dab had carried it off with him, also the poultry house key. They were tied together. I told Gibbie the keys were not in their usual place, and asked if he could manage to get in the stable without breaking the lock.

How to lay the breakfast table.

He answered that was very easy, and proceeded to roll the wagon with the rack up to the stable door, climbed into the loft with ease, and thence down the ladder into the stable, where he unbarred the back door. I was a most interested spectator, for now I understand how the horses are ridden at night, when the door is locked and the key hanging on its hook in the pantry. I did not tell Gibbie that Dab had the keys; I preferred to let him think I had mislaid them.

By the mail at 11 o'clock came a postal addressed to me with the keys attached. I am truly glad Dab had the decency to send them.

April 7.

A long and humble letter from Dab making his apologies as best he could for his very bad conduct and thanking me for all I had done for him and saying he had no fault to find with me, that he knew if he had come to tell me he would never have gone, intimating that it was with members of his own race he found it hard to get on; said he saw Rab every day and his hand was getting better. He wound up by begging me to give him a recommendation.

Meantime Patty is covering herself with honors and we are getting on very well.

My strawberries are fine, we are picking four quarts every day. Green peas are also bearing well. It is a great thing to have them now while the house is full. We had the first strawberries on April 11.

Rained hard all night, but cleared beautifully this morning; such a blessing to the young corn the rain was. L. wanted to see Casa Bianca, so we drove down there and had a delightful day. We got back in nice time, but Gibbie had gone home, so I had to take out the horses and then go down to the barn-yard to get out feed, as Bonaparte is still working on the house at Peaceville. I miss Dab terribly; he was so quick and always so ready to do everything.

L. went with me to get out the feed. She has lived in a city always, and it must all seem very strange to her. We counted out eighty-four ears of corn into the sack, and then the problem of getting it moved came up. It was still raining, and the horses were eagerly following us, almost walking over us.

L. kept them off with the lantern, while I attempted to drag the corn along. Just then Gibbie strolled up, to my great relief.

April 8.

Drove L. to Gregory to take train. She has kindly offered to take charge of Rab on his journey. I went to ask Dr. G. if Rab's hand was in condition for him to go. He said it was, that it had healed very rapidly, being perfectly healthy, and no longer needed to be dressed daily.

I think it wise to send Rab back to the institution of the worthy Jenkins. I dare not leave him at home with Chloe when I go away, as I must do in ten days to be absent a month, for there is no telling what he might do. As Chloe expresses it, he "discounts her altogether," and this craze he has for firearms makes her afraid to keep him. Under Jenkins's charge he will be well cared for, and at the same time kept out of mischief and made to behave.

I wrote to Jim yesterday to have Rab ready, as I would call for him this afternoon, and as I drove up Hattie came out with Rab's valise, and he followed with his arm in a sling, but looking much better. We drove rapidly to the train, just in time to get the tickets and get L. and her protégé on the train, before it was off.

I asked Rab as we drove down if he had seen Dab. He said he had very often, that he had got a place as butler, where he was getting $5 a week. I asked where it was, and after the train left, I drove to the house and asked to see Dab.

I told him I could not give him the recommendation I had expected to give him, because he had run away and left me as he had done; that I only wanted to see him to tell him to keep the place he had, and not to run from place to place. He seemed much moved, and so was I. I sat in the buckboard, and he stood by the hind wheel, so that I had to turn to look at him.

I gave him a little lecture, telling him that I had carried out my promise to his dying mother as far as I could, having taken much trouble with him, as well as being put to a good deal of expense, because of that promise, and that now he had taken the matter out of my hands by leaving me. All I demanded of him was that he should lead a respectable life and be industrious, honest, and upright, and I would be satisfied. When I turned to look at him the tears were rolling down his cheeks and he thanked me and said he would try.

I started on my lonely drive of fourteen miles about 6 o'clock.

My thoughts were ample company, for I have much to plan out.

All winter I have been looking forward with great pleasure to a visit from a charming English friend who stayed with me once a few years ago. She has made a trip around the world with her maid and physician and was coming here on her way from San Francisco to New York, but after a visit in Mexico, for some reason the physician thought it would be unwise for her to come to this remote plantation, so far from railroad, telegraph, and I suppose he thought from civilization. Mrs. R. wrote to tell me of her disappointment at this and to ask me to make her a visit in New York instead, and begging me to bring Chloe with me. This royally generous invitation I have accepted, and my mind is much occupied as how to arrange for the care of everything in the absence of two such important people as Chloe and myself.

CHAPTER X

April 9.

MY wedding day thirty-six years ago! It does not seem possible that there can be one atom of the intensely pleasure loving, gay slip of a girl left in the philosopher who, battered and bruised by life's battle, looks with calm, serene eyes on the stormy path behind her and with absolute faith forward to the sunset hour. It does not seem as though the ego could possibly be the same. Had some magic mirror been possible, in which that girl could have been shown herself, and her solitary life at the end of forty years, she could not have faced life, she would have prayed passionately for death.

Everything she specially cared for and valued has been taken from her, the things she specially disliked and feared have come upon her, and yet all that is great and noble in life, seems nearer to her now. God seems to have turned all the evil into good, all the mud and mire into gold, and there are around her the beautiful mists and clouds of the sunset, which is not so far off now. So does the Great Father fuse and mould and change in His mighty workshop. Thank God for His alchemy.

April 10.

Spent the day at Casa Bianca sheep trading. I am no trader and should have some one else to do these things. I am always afraid of taking advantage of other people, and as a consequence I am generally a severe loser.

My sheep are fat and have not been shorn and they have been a paying investment, the best I have ever had, but they are being stolen steadily. Last Wednesday we counted twenty-four sheep and fifteen lambs, and to-day I could only count twenty-three sheep, and this has been going on a long time.

There is one splendid ram and the lambs are beauties, but Capt. M. only paid me $54.75 for the whole lot. I also sold two cows which I was still milking for $10 apiece. I need the money and have to take what is offered.

April 11.

Sent Gibbie yesterday to take the two cows I sold down to the ferry. The cows are very gentle so that I never thought of any trouble. In the afternoon went out to ask him about it. When I asked what time he reached the ferry he seemed much embarrassed, scratched his head and stood on one foot and then the other and finally said he never got down to the ferry. He stopped to talk to some one and the cows were eating, and the first thing he knew they had got away in the woods, and he had of course pursued them with great activity, but to no purpose, and finally gave it up, and when he got back home found them waiting at the gate.

So that has all to be done over, and I have to write and appoint another day for them to be met at the ferry. It is very discouraging. Nothing that I cannot personally attend to gets done. Poor dear old Bonaparte cannot help; he can only denounce and condemn "this new giniration," which does no good at all.

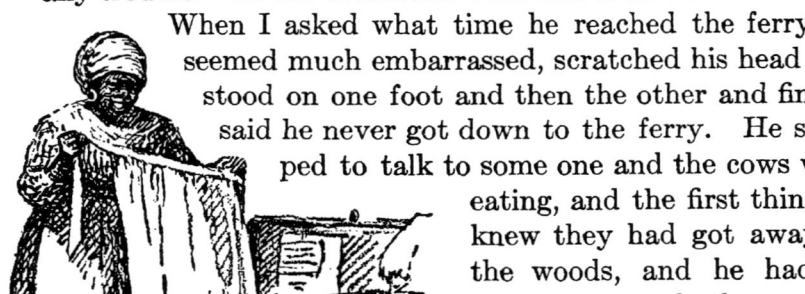

Joy unspeakable.

CHEROKEE, April 12.

Such intense excitement pervades this household that it is difficult to accomplish anything. The last two weeks have been very full. Corn has been planted, also potatoes, and land prepared for cotton. The incubator hatched out a splendid lot of healthy chicks.

Besides all this I have been sewing and dressmaking, for the day after to-morrow I am leaving for a visit to New York, and, wonder of wonders, Chloe is to go too by special invitation. I was afraid at first the excitement would put an end to her, for when I read the letter of invitation she seemed overcome.

At first she said it would be impossible for her to leave the chickens, and who would take care of the house and yard while she was gone? No, it was impossible. But I arranged to get Jim's wife to take charge of the precious "'cubators," also the whole poultry yard, and Chloe is to go. She prides herself on being a travelled person, having been in North Carolina and Georgia, as well as to many different parts of South Carolina and having gone all through the public buildings in the capital of this State, but the idea of going to New York and having to pass through Washington going and coming — it seems too much.

Besides this a complete outfit had to be got for the journey. That of itself was joy unspeakable. My own preparations sink into insignificance beside the magnitude of those of my good Chloe.

April 13.

We drove to Casa Bianca, where we had lunch, and M. and L. left us and drove to Gregory to take the train. It had been an ideal day.

Told Nat he must come to Cherokee to-morrow and drive down a bunch of young cattle, as the pasture there is splendid and I have only two cows, while at Cherokee the pasture is poor and I have twenty-four head of cattle. Nat said he could not possibly bring the young cattle down, that they had never been outside of the enclosure at Cherokee and that as soon as they got out they would all scatter in the woods and be lost.

He has always been a good hand with cattle and has three cows and calves and a pair of oxen at Casa Bianca, as sleek and fat as possible. I was surprised at his refusal and told him he could get one, or even two, boys to help him and I would pay for it, but still he insisted he could not do it.

At last I said, "If you have any trouble in getting off with them, I will get on my horse and help drive them myself." At once he brightened up and said: "Bery well, miss, I'll cum for dem to-morrer." His refusal and the consequent discussion delayed us greatly and we were very late getting home.

April 14.

Yesterday at nine Nat came for the cattle. I went out and had the good Martha, who is as quiet as a cow can be, roped so as to act as pioneer in conducting the others down. I am milking her and am sorry to send her away, but she was born and reared at Casa Bianca and is always overjoyed to go there, so that nothing will make her leave the road.

Equinox, the beautiful young bull, with John Smith, the two-year-old steer, and Ideala, a beauty three years old, and Pocahontas, Virginia, and Queenie were the party. They are all very gentle and started out of the front gate quietly and I returned to the house, but before I had taken my seat at the sewing machine Nat sent for me.

"Miss," he said, "yu know yu promise yu go too en help me."

"Oh, Nat, that was only in case there was any trouble. I only said that for fun. I knew they would not give any trouble. See how quietly they went out of the gate."

"But, miss, I neber would 'a' cum ef yu neber say so,'case I know dem cows gwine loss. En, miss, yu done promise."

They know a promise is sacred with me. There was nothing for it but to tell Green to put the saddle on Romola at once, and to prepare for the sixteen-mile ride.

It was very provoking. I had some sewing which was most important, and I have so few days at home and with D. now, but I told Nat to go on with the cattle and I would catch up with them in a few moments.

Green is slow about saddling, so they had gone about half a mile beyond the avenue gate when I came upon Nat alone in the road with Martha, who was going round and round, while Nat used his long lash upon her. I called to him to stop at once, and I asked where the others were. He answered in an I-told-you-so voice: —

"In de 'oods, en Mahta want to git dey too."

He had sent the two boys after the young cattle instead of tying Martha to a tree and going too. I said nothing, but rode out into the woods, and after some little trouble brought them back into the road, where by great vigilance and activity we managed to keep them.

As they were unaccustomed to travel they went very slowly and we had often to stop in shady places and let them rest. When they came to a stream of water crossing the road they would lie down, and there was nothing but wait.

However, all the irritation of giving up my plans at home passed from me and I soon was thoroughly enjoying God's beautiful world — the fresh air, the lovely wild flowers, the birds and bees, all rejoicing in the return of spring with its promise of fruition — it was all a joy.

At last we got the party safely through the gate at Casa Bianca, and when they came to the turn where the avenue runs along the river they felt rewarded for all their trials, such thick, rich grass under their feet, cool shade above them, and that great stream of water beside them.

I had not brought the house key, to my sorrow, for there

I keep a demijohn of artesian water and a box of crackers — the well is so little used that I do not like to drink the water. I turned my face homeward very hungry and very thirsty. As I rode down the avenue I saw a great mulberry tree loaded with ripe fruit. With delight I rode under the branches and satisfied both hunger and thirst and went on my way refreshed.

It had taken so long to get the cattle down there, what with the various stoppages, that it was 5 o'clock when I got home. Found D. had been very anxious about me and was much surprised to see me so little exhausted by the day. Chloe gave us a delicious dinner and I was not too tired to walk over the fields with Bonaparte and give him directions for his guidance during my approaching absence.

Everything is now about ready and D. V. we leave here next Wednesday. I never can keep up my diary while away and will not attempt it.

I have taken a little boy of 8, Elihu's son, to take the place which Rab and Dab have successively occupied for years about the yard. I cannot afford to keep a man-servant at the pineland. This little boy's name is Green, but he is so strong and capable that I call him Goliah. He did not like it at first, not until I told him Goliah was a giant. I asked him if he had no nickname, as I never could remember to call him Green. He answered gravely that he had a nickname, and when I asked what it was he said "Isaiah." A most unusual nickname, but it seemed to open the way for me, so I said: —

"My nickname for you will be Goliah, because you are so strong."

Poor little Goliah was in rags and I have made him some clothes, but my forte is not tailoring and I could not get just the stuff I wanted for him. He speaks of himself as my "'ostler." I speak of him as my "man of all work," for such he is.

Sunday, April 15.

After service I went over to my house in Peaceville, which is just opposite the church, and took out four queer little, old-fashioned trunks full of papers which I have kept out there until now. Two of the trunks are covered with skins with the hair on and studded with brass nails. One has the initials "E. F. B." in brass on the top.

They all contain very old papers, among them grants to my ancestors for 6000 acres of land. These are very much the worse for age and I am going to take them on to Washington to see if I can have them repaired. I scarcely

The church in Peaceville.

think it will be possible as the grant to my great-great-grandmother, Esther Allston, is falling to pieces, and the seal seems in danger of crumbling. The date is December 21, 1769.

I did not know the grants were in these old trunks, which I was gradually looking over. I kept them at Peaceville because the summer days are longer and more suitable for reading old letters and papers. I have been urged by two publishers to write all I can remember from my earliest years. It seems to me absurd for one who has lived such a secluded life to write her reminiscences, but I would find it most interesting work, as it would involve the reading over of old letters. I have every letter written to me since I was 10 years old. If the pressure of daily anxiety for the wherewithal to carry on the work is ever lightened I think I will try to do it just for my own satisfaction, for I do not think it would ever be a profitable venture for publication.

I have always kept a diary of some sort. When I was

married and was ambitious to become a fine housekeeper, though I could never hope to rival my belle-mère, who had a genius for housekeeping besides being a brilliantly clever woman, I kept a "Diary of Dinners," in which I recorded every culinary triumph of my belle-mère, with whom we lived for two years. Then when thrown upon my own resources, I had this delightful guide to the possibilities of the season as to dinners.

It was not so difficult then to provide because we raised great quantities of poultry, turkeys, and ducks and guinea fowl as well as chickens, for the negroes did not steal things then as they do now; they all raised an abundance of poultry themselves and so the temptation to steal was not so great. Now they raise less and less poultry every year. This comes from their selling all their chickens and eggs and buying canned salmon, sardines, biscuit, and ginger snaps.

April 16.

Left home at 11:30, drove to Woodstock for luncheon with my brother and then on to the station. There the two charming little travelling mates I am to have met me. Son is $4\frac{1}{2}$ and Sister $2\frac{1}{2}$. Their fair hair, lovely brown eyes, and piquant little retroussé noses were a joy to watch. Everything interested them; nothing escaped them. Their father went with us as far as Lanes, where we changed cars and took the sleeper.

I thought this would cause a breakdown and tears when he left them, but there was none. He had provided them with a liberal supply of bananas and candy, which rather alarmed me, but occupied their full attention, and with the wonders of the transforming of the seats into accommodations for the night there was no space for homesickness or sadness. When I proposed bed they were eagerly acquiescent, and Son was most efficient in producing all that was necessary from the tightly packed valise.

The greatest problem was to get off Sister's little dress. The dear mother wishing to save me trouble had made the little travelling frock of a pattern which called for no buttons; it was simply slipped over the head, but it was so close a fit that it seemed to me if I pulled it rashly off, Sister's dear little tip tilted nose would go with it. Son at last said: "Now, Sister, don't cry; I just have to give it a jerk over your nose; it will hurt some, but not too much."

So Sister braced herself to stand the jerk and off it came, leaving her little pansy face unhurt but very rosy. I tucked them into the lower berth, opposite mine, and after a few suppressed ripples of laughter have not heard a sound from them.

April 17.

We reached Washington on time. The dear little children slept like tops all night. I woke them at 7. Son again proved himself a most accomplished nurse-maid, and Sister emerged from the train looking very dainty and fresh. Son insisted on struggling to carry their heavy valise, but was finally persuaded to let the porter take it with mine out to the gate, where my dear sister was waiting, and the children uttered a cry of delight as they recognized their beautiful aunt with her husband, their unknown uncle, who had come over from Philadelphia to meet them. We parted company here, and I could truthfully say they had not given me the smallest trouble, but on the contrary, had been a genuine pleasure.

THE CAMPS, June 11.

I left Carollton on the 4 P.M. train. En route anxiety came to me as to whether I had given my letter, telling when I would reach Gregory, time enough to precede me. As I neared my destination I felt more and more sure that I had not. If I had mailed it myself it might have arrived, but I gave it to the children who were playing in the garden and asked them to drop it in the box. A child's hand is almost as dangerous a place for a letter as a man's pocket, and I had an inward conviction there would be no one to meet me at the train.

I asked the negro porter to look out and see if the phaeton was there to meet me when we arrived at 10 P.M. He came back and told me he did not see it. Then I asked him to engage a hack to take me out the three miles to this pineland, where I was to spend the night and make a little visit to C. By the time he had made sure there was no one to meet me and reported this to me, every vehicle had gone except a huge omnibus with a large pair of mules driven by a small darky who looked about 10. He was eager to undertake to get me out to the Camps for a small sum.

"Do you know the way?" I asked.

"Yes'um. Oh, yes'um; know um well."

So I cl mbed into my chariot, where a feeble lantern hung. A still smaller urchin slammed the door, and I started. I must say I felt I was doing a rash thing, for I was not at all familiar with the road myself, and by this time it was 11 o'clock. As long as we were within the radius of the electric lights of the town I didn't feel so anxious, but when we got into the blackness of darkness I began to think how foolish I was not to have gone to a hotel for the night.

Every now and then my Jehu would climb up to the front window, where I stood peering out into the night, and ask, "You t'ink we git dey yit?" I could faintly make out houses at intervals along the way on each side and was sure we still had a long way to go. At last when we got into a denser growth of pines and I could see nothing I called to him: "Stop, and I will walk the rest of the way if you will bring the lantern!"

Greatly relieved, I think, for he began to fear he was to drive all night, he got down, charged the other mite not to let the mules stir, a command which seemed to me superfluous, for they were only too glad to stop, and with much difficulty undid the wire which held the lantern in place, and we started. I knew that if we had come to the place to turn in to C.'s house there was a narrow bridge with a sharp turn which it would be difficult for the very large vehicle and mules to make safely. After a little wandering around, most of the houses being in darkness, I saw a light in a house, and as we approached the fence, the dogs gave tongue and I knew I was getting to the right place. The dogs are fierce, so I stood and called for some one to come, rejoicing that this family were not as early in their habits as their neighbors.

My little guide now resumed his confident air and said: "I t'ink, ma'am, you ought to pay me mo'n I charge you fust time."

"Oh, boy," I said, "for this voyage of discovery I will pay you double what you charged; here it is. Now, tell me, were you ever here before?"

"Not to dis place, ma'am, but onst las' year I bin about halfway here, but I didn't bin a dribe, I bin on me foot."

I felt that the Providence which is said specially to protect fools and children had been with us. I felt anxious as to how the house on wheels was to be turned in the narrow road with a deep ditch on each side, and proceeded to offer some suggestions, but this individual of resources stopped me by saying, "Needn't fret, ma'am. I onderstand dribe," and I was free to enjoy the pleasant welcome that awaited me within. C. had received no letter from me and had no idea I was coming.

PEACEVILLE, June 14.

Got back from my delightful holiday last evening. I stopped on my drive from the Camps at Cherokee to see how everything was. Found my good old Bonaparte in deep distress; his faithful and devoted wife died two weeks ago.

They had lived together happily fifty-three years and he is crushed. He cried like a child on seeing me. I gave him my earnest sympathy. I'm so sorry her illness and death should have come while I was away. In broken words he told me what a surprise it was to him; he never thought "Liz could die en lef' him." When a working-man loses a good wife he is indeed bereft; his companion, helpmeet, cook, washer, seamstress, mender, all gone at one fell swoop, and he is left forlorn.

I shall myself miss Lizette very much. There were certain things she always did in the sausage making and Christmas preparations. I always meant to get her to tell me all she remembered and to write it down. She belonged to a family much considered by my father and by his parents before him. They were distinguished for loyalty, fidelity, and honesty, and took great pride in their distinction as a family.

Lizette's mother, Maum Maria, was our nurse and her father, old Daddy Moses, could be trusted with anything. Put gold, silver, provisions, anything in his charge and it was safe. His sense of responsibility was sacred — alas, alas, to find such a one now! Some of his descendants are very smart, but none has just his character.

One of his sons, William Baron, who had been our third house servant, I mean in rank, there being two men above him, made quite a name for himself as a caterer and steward of the club in Charleston after the war, and one of his great-grandsons Sam Grice (Lizette and Bonaparte's grandson) is a minister of the Episcopal Church. He was educated in a church school here and when there was a call for a boy of high character to be taken and to be educated by the church he was chosen and proved most satisfactory in every way, and he passed a remarkably fine examination in Greek, Latin, and Hebrew before his ordination.

He used to visit his grandparents every year, but since he

has been ordained priest he has not been here, as he has married and has work. He is the pride and joy of his grandparents' heart, but there is always a little drawback in the fact that they are uncompromising Methodists and they feel he has deserted their church.

For the present Bonaparte's usefulness is quite gone, even his capacity to do anything with his hands, and he cannot stay in his house, has to go over to stay at his son's, a mile away. He seemed to feel that now I had come home things would be easier.

Chloe came home two weeks ahead of me so as to accomplish the move from the plantation to the pineland, and I found everything comfortably arranged for me, a nice dinner and no bad news.

My delightful new possession, a most high-bred and distinguished Scottish terrier, MacDuff, which I sent home with her, met me with enthusiasm. He was a present from my charming friend and hostess and is going to be a great pleasure to me. It is impossible to describe him, he reminds one of so many different wild beasts, all the time being strangely human.

After dinner Chloe brought out a beautiful fruit-cake which she had made for my birthday. She seemed afraid I might accuse her of extravagance and assured me she had only used up the odds and ends of fruit which were left in the storeroom and the fresh butter and fresh eggs which I was not at home to eat. I was delighted and praised her very much for her cleverness and thought; seeing me looking in the silver drawer for the cake knife, she added hastily:—

"Mind, Miss Pashuns, I ain't tell yu fu' cut um till yu hab kump'ny."

So I said, "That is very wise, Chloe; put it away until we have company," and she removed it with great agility, but it was a disappointment, for I have got accustomed to having

z

a great many nice things all the time recently. I know it is going to be hard to force myself back to the great economy I have felt necessary and practised for the past year.

June 16.

Rose at 5 : 20 o'clock and had breakfast early. That is one of the unexpected results of Chloe's travel; she is much earlier in the morning, which is a great comfort; that is the only cool time, and I am so anxious for Jim to get off to his work at 6 : 30 o'clock every day; it is much better for man and beast to start the ploughing very early, and then knock off for the hottest hours and plough again in the afternoon.

The season has been hard on all crops; a severe drought after the late frost, so that it was hard for seeds to come up. I have nice snap beans and corn from the garden and soon will have tomatoes. The cotton and corn in the field look poorly, the watermelons need work, but I hope they will be abundant.

Chloe's visit to New York is a subject of immense and unending interest to every one. She spends her time narrating to white and black all she has seen. She brought most carefully selected little presents for every one. How she managed I do not know.

Chloe was a great success at the North.

The truth is Chloe was a great success at the North; the height of her white turban, the width and length of her white apron, the classically disposed white kerchief crossed over her ample form, the large gold hoop earrings and her Mona Lisa smile as she dropped a curtsy to any guest appearing at the door of my sitting room at the St. Regis impressed those unaccustomed to it very much.

Her ready answers to all questions were most discreet. A friend of mine asking her what she had seen in her short stop in Washington said, "Did you see the President?"

"No, ma'am, I ain't see de Presidence, but I see de gold pianner," that piece of furniture of the White House seeming the full equal in interest and grandeur of the head of the nation.

Chloe's face during Buffalo Bill's Wild West Show was a study. She would not give way to surprise of any sort, but occasionally I felt a violent punch in my back when Chloe's excitement had reached a point where some action was necessary and she was afraid I might miss something — we were in a box. The presence of none but white servants was very unexpected and unaccountable to Chloe, but she made no sign. She spoke with pride of the table she had to herself and how attentive every one was. She said: —

"Miss Pashuns, I never hurry fu' eat. I look 'roun' en enjoy meself. Fust thing I had fu' brekfust, I had a oringe. I jes' wait en res' meself till I see de lady to de nex' table cut she oringe een half en tek de spoon en eat um wid de spoon, den I dun de same, but I neber let um see I watch um.

"Den de gentleman tek dat plate way, en bring some hom'ny een a saucer. Den I watch de lady en see um put shuger on de hom'ny en por milk on, en I done de same. Den de gen'leman tek dat 'way en bring me sum aig, but I tell um 'Thank yo', sah, but yu needn't truble yo'self to bring me no aig, kase I don't eat aig, neither no mutton kase I don't eat dat needer.' I didn't like him to hav' de trubble fo' bring um en tek um back."

The second day she was there she was quite agitated.

"Miss Pashuns," she said, "I 'most had a accidence. W'en I git een de allivatu de nyung man staat off mos' too quick, un lik' to t'row me down, en 'e was dat skeer till 'e trimble en 'e ketch me a'm en 'e say 'Is yu hurt?' en I mek ansuh,

'No, sa, I ain't hurt,' den 'e say, 'Please don't tell no one. I hope yu ain't hurt, fu' dat would git me in big trubble.' Den I promise I wouldn't tell.''

On the way home Chloe stopped several days with my sister in Washington, who took her all over the public buildings. She saw a great deal that I never have seen because I always have so many other things I want to do while in Washington.

When she was being taken through the Capitol and saw in the great hall the statues of distinguished men she went round and examined each one very carefully, then came to where L. was sitting waiting for her, and said in a low tone very wistfully: —

"Miss Luise, Ole Maussa ain't yere."

When L. answered: "No, Chloe, papa's statue is not here," she heaved a sigh of deep disappointment.

"Ole Maussa" to Chloe was the greatest man in the world and she thought less of the Capitol when she did not find him.

When any one treats her with scant courtesy or intrudes on her feelings in any way she is in the habit of explaining: "My master was de Guv'ner en I kno' how tu behave." We showed her the family name in the ceiling of the beautiful library building, telling her it was the name of Uncle Washington, whose bust was in the dining room at home. After craning her neck for a long time her small book-learning enabled her to make it out for herself and she was greatly pleased.

Chloe has been made very proud and happy by the graduation of her granddaughter Clara with great éclat. She is only 16 and very small and childish looking, but she took her diploma and made a very fine speech. Chloe told me, when the principal came to speak he said: —

"For five years the name at the head of every class she was in, was Clara Galant and not a black mark against it."

"She was dressed very fine een a w'ite silk net, a twenty-dollar frock, en w'ite shoes en a big w'ite bow on 'e head, en everybody say 'e speak butiful en dem was surprise." I was greatly surprised to hear all this and very much pleased. Bonaparte has a grandson who has distinguished himself, passing the test examinations in Latin, Greek, and Hebrew and is a minister of the Episcopal Church and is now in Virginia. That has given me much pleasure, and now to have Chloe's little granddaughter distinguish herself is very nice; and dear, faithful Chloe is so proud and happy. Clara received very handsome presents from people in Gregory, two gold pieces of $10 each, a silver set of writing implements, and many more. I suppose people were anxious to show their appreciation of her faithful good work in school.

The house looks so fresh and clean in its new coat of whitewash and feels so solid and unshakable after its thorough repairing that I feel as though it was a palace. My dear Chloe has brought out all the pictures and books she thinks I would like to have. Her selections always amuse me. "Forty Days of Lent" is one book prominent, and the King and Queen of Spain in bridal array hold the place of honor over the mantelpiece! After all it is good to get home, though it may not be a bed of roses; there dwell your Lares and Penates, and there only. Jim reported Chloe's other granddaughter, Josephine, as very ill; she has a baby three weeks old. I told Chloe she must go down at once. She began to say, "Impossible to lef' yu, Miss Pashuns, wid nobuddy but dis gal." But I would not listen. I ordered Jim to put Ruth in the buckboard at once and told Goliah to make himself decent to drive her.

It was impossible to get her off till after 5. I fear from what Jim says there is no hope for Josephine. He said they were giving her an ice bath when he left she was so burned up with fever.

I took MacDuff to sleep in the house as there was no one anywhere near the house. I have been practising a good deal lately and to-night played for two hours.

June 20.

A quiet night alone with MacDuff. He behaved very well, though he does not like staying in the house as Mops used to do. When I tried him before he walked about all night, making so much noise that I could not sleep.

Lizette cooked some hominy and corn-bread very well and boiled an egg, so that I had a good breakfast. Chloe returned at 8 o'clock, just as I had finished. Poor little Josephine died ten minutes before she got there, but she had the satisfaction of sitting up with the body last night, and left early this morning to give orders here about digging the grave, as she begged them to "bring her home en put her by her mudder." She told me the "castle" was ordered very fine and that she was beautifully dressed. She was wonderfully composed. I told her to lie down and rest at once.

Then she confided to me that she had nothing suitable to wear at the funeral. Nothing black but a silk trimmed with lace. I went and ransacked all my belongings and at last found something that I thought would do. Unfortunately Chloe is formed in a more generous mould and the present cut of skirts makes it difficult to stretch them, but Jim's wife, Hetty, happened to be here and she is clever with her needle, so she undertook to enlarge the skirt, while I got a black hat and trimmed it and found a suitable veil, so that by the time the funeral procession arrived from Gregory, Chloe looked very nice.

To-night before going to bed she gave me an account of it with great pride. The "castle" was beautiful and four carriages and three buggies came up from Gregory behind the "hurst." One of Josephine's aunts has adopted the baby. I wanted Chloe to take it, but she does not care for children.

June 21.

A most exhausting day. The only way we have of making money for our auxiliary is by making ice-cream for sale two or three times during the summer. Ice-cream is a rarity in Peaceville and consequently these sales are very successful and we had arranged to have one this afternoon. I and my dear little neighbor, who is secretary of the auxiliary, furnish part of the milk. As soon as I could get off after the many impediments which arose this morning I took the demijohn of milk and drove over to Mr. F.'s and got theirs and took them out to Peaceville, where Mrs. R. and J. F. are going to make the cream.

I came home, had a hurried dinner, and went back to Peaceville to serve the cream, which I always enjoy, but the heat and the drive back and forth, amounting to sixteen miles, were almost too much for me. I brought some ice-cream to poor ill Georgie. I had the can packed well in ice and her delight over it was pathetic. She washed off the salt and ate all the ice after finishing the cream. I also brought some for Chloe and Lizette.

When I got seated down in the cool dining room in Mama's big chair and my little lamp with the shade on it I was too tired to move until after 12 o'clock. On the table by me was a book which I read in every spare moment with much pleasure: "The Bible in Spain; or the Journeys, Adventures and Imprisonments of an Englishman in an Attempt to Circulate the Scriptures in the Peninsula." George Borrow made these journeys as far back as 1835, so there is nothing new in the book, but it holds my attention when I am too tired to read anything else, and to-night it did not fail me.

June 22.

The cotton is coming up, also the corn which was so long in the ground. I am so glad cow-peas are selling for $3 a bushel, and I am having mine threshed out so that I can sell some and be able to pay for the hoeing which is absolutely necessary now. As long as the cotton had not come up it seemed dangerous to attempt to work it.

Jim is in great distress because the doctor says his little girl has tuberculosis and that unless she is brought into the country and kept out of doors she will not live until August. He wants to break up in town and move into the country, but the wife will not. As Jim says, "Seems like they rather die in town than live in the country." So he asked my permission to bring her up to stay with him. Of course I consented.

Chloe came to tell me she had got a letter from her sister saying she must go down to-morrow and take $30 with her to pay for the funeral expenses. I said: "You have that much in the bank, Chloe?" She said yes, but after a while it came out that she had taken all her money out of the bank at her sister's bidding to buy finery for Clara to graduate in.

I was quite distracted, for I will have to borrow the $30 to give her, and I never know where to borrow money. I once borrowed $1000 from the bank. It was when I was planting rice successfully and had no doubt as to paying it easily when the crop came in. But that year some misfortune happened and I thought I should lose my mind over that debt. I had given a mortgage on Casa Bianca. It was a year of great depression in this country from loss of crops and the low price of rice, and if there had been a forced sale the place would have gone for nothing. Since then I have done anything rather than borrow — but now for my dear Chloe I must do it.

July 4.

A brilliant day for the darkies to celebrate; it is the day of days to them. Lizette has been in such an excitement that she broke the top of one of my precious little pink Wedgwood dishes. I could have cried if I had not been ashamed; having no people around me I get so fond of things. Goliah behaved abominably, refusing to crack the corn for Chloe before he went off, though I had given him 10 cents and a watermelon.

Had dinner at 12 so that the servants could all go and had a most delightful long afternoon. I took my sewing and book and sat down by the river with the dogs. When I found it too dark to see either to read or to sew I chained Don and then came in and lighted the lamps and had my tea.

Chloe returned about 10 o'clock. I had sent poor little Georgie a present of a melon by her, and she said: —

"Miss Pashuns, ef yu cud a see Georgie w'en I g'en she de melun! 'Twas teching! 'E say 'e had a dreem 'bout mellun en dem so scarse. Moses cudn't give him money f'r buy none, en now 'e hab one, en 'e say 'e cudn't tenk yu 'nuff."

July 11.

S. came up and made me a delightful visit.

Though there is a great gap of years between S. and myself we have so many of the same tastes and interests that the years do not count in our intercourse. Her music is a delight to me, and it is such a wonder that she keeps it up as she does with so many drawbacks and with such an old and weary piano. I often feel that I would like to give her my Steinway, which, when I come to count the years, is itself not in its first bloom, having been bought in 1885; but it is an infant compared to hers and would be a joy to her, the action is so good and the tone so full; but really I would not dare to face my existence here without it. I shudder at the thought; so I hastily quench the impulse.

This afternoon I brought back seven nice watermelons from the plantation, greatly to Goliah's delight. They weighed down the buckboard so that he proposed to walk home to lighten the buggy. I suppose he weighs about fifty-five pounds. I thanked him for the proposal, but said I did not wish to reach home before him. Oh, no, he said, he would run and keep up; but I would not let him.

Little Goliah is the happiest, jolliest little boy, so fat and so black and shiny. My efforts to teach him are futile in the extreme, but why should Goliah be taught anything? He has a vast fund of general information of things to me unknown, and above all he has such a power of observation that nothing escapes him.

I am absent-minded and constantly lose keys and things like pencils and handkerchiefs, etc. When I ask Goliah as to what I had in my hand when I spoke to him last he can always tell me accurately, and my next question is, "And where did I go when I finished talking with you?" He can always tell exactly, and, moreover, I always find that he knows every step I have taken since, though he is in the yard and I am in the house.

If I say, "Goliah, remind me to-morrow to write a particular letter" or to do any special thing, he is sure to remind me. He has learned to wash his clothes so beautifully white that it is a pleasure to see him — to all but Gibbie, who is very much provoked at Goliah's white suits, only varied by a sky blue suit. He grumbles aloud, and I heard him say, "Miss Pashuns hab dat chile dress up all een w'ite till 'e far' look like a shadder; 'e skare me."

Altogether I consider Goliah a luxury. I have not the luxury of electric lights nor telephone nor automobile nor ice, but I have unlimited space and fresh air and sunshine and the wild flowers springing up everywhere around me, and this little piece of animated nature just bubbling over with life and joy and the absolute delight of having plenty to eat and nice clothes to wear and being always clean and owning a spelling book and slate and a bed of his own and a little trunk, also saying a very mild lesson every day and catechism on Sundays — all these things which to most children are a matter of course are to him something quite new in his little experience and pure bliss.

When you add to this that he has Ruth, that big fiery looking animal at his command, and that when he has been out on Sunday to visit his family and appears at the gate on his return she whinnies and goes to meet him, really his little cup, for eight years empty, is full to overflowing — and what gives me so much pleasure is there is no arrière pensée, no déjà-connu — it is all so fresh and so perfectly natural. Of course I know it cannot last.

Goliah is a constant amusement to me. I am teaching him to drive, and I read, for when it is very hot and the horse seems to feel it as much as I do, I cannot make her go fast, and I get so impatient and so hot that it is an immense relief to have a magazine to read. Of course I have to keep an eye always on Goliah and the reins. He stands at the back of the buckboard, finding that gives more power than sitting.

He talks constantly. I think he conceives it part of his duty to entertain me. "You see dat bu'd, Miss Pashuns?" A large brown bird which would light in the road and when Ruth got within six feet of it would fly, to light a little way ahead, waiting until I thought the horse must tread on it.

"Yes, I see the bird."

"Yu kno' wha da bird does say? 'E tell eberybody, 'Plant bittle fu' winta! Plant bittle fu' winta!'"

Now we call the bird a chick-will-willo; it is a first cousin of the whip-poor-will and has a more cheerful note, but I had never heard any sense attributed to its incessant and insistent note before, and I was delighted with the darky version.

"Oh, Goliah," I said, "what a pity people will not mind him, there is so much land and so many idle people; if they only would plant victuals for winter what abundance there would be for everybody, man and beast."

At which he informed me that he had planted a corn crop himself before he came to me and as we passed his father's house he showed it to me with pride, the feeblest growth.

I am trying to teach him to read and I'm sure it would be easy if he could only learn his letters, but I cannot accomplish that. He says the alphabet off glibly, but the letters seem to look all alike to him and my efforts to describe them don't seem successful. I point to a letter and say that is T, it stands for "table" and looks like this table — showing him one with a leg in the middle and two leaves. S stands for "snake" and looks like one. A is like the step-ladder. But when I go over them he knows not one, unless he says the whole alphabet and stops at the letter. I try making him copy the letters on the slate, but nothing seems to impress them on him; and yet he is so clever in learning his catechism and hymns.

This village feels it has taken an immense step forward since the honk honk of the automobile can be heard here daily. Fortunately the owner is very considerate of horses and slows down and even stops if necessary, so that Ruth is getting quite over her fright about it. All she wanted was to understand what it was, and now she is beginning to recognize it as a new kind of horse of great speed. When it passes her on the road she tries her best to catch up with it.

CHAPTER XI

PEACEVILLE, July 7.

IT has been desperately hot and when I got a cordial invitation from Mrs. G. to spend a few days with her on Pawleys Island I was overjoyed. My old summer home was there, and since we had to sell the place ten years ago I have never been willing to see the beach again, but now I am

My old summer home at Pawleys Island.

just gasping for a breath of the sea and I made my arrangements to go to-day.

I had Jerry King ploughing in cow-peas at Cherokee, and he is a fine boatman, so I told Bonaparte to have my little dugout canoe which I call the Whiting ready for me at the wharf at 10 o'clock, with Jerry to row me. When I drove down, what was my dismay to find no Jerry there. Bonaparte with unmoved dignity told me that Jerry had just been arrested by the Sheriff while ploughing in the field, for debt, he said.

I was quite distressed. Jerry does not live on my place and so I know nothing about his financial status. I had to find some one, for Mrs. G.'s surrey would be waiting for me on the other side. Jim was eager to row me, but I doubt his being able to hold out for a seven-mile row, not having used oars for years. I found Aaron was working his cotton in the far field, so I sent for him.

He was taken unawares and came not knowing what I wanted, and was most reluctant to go without being dressed for the occasion. However, I insisted that it was an emergency and he would have to forget the good clothes he would like to have on, and I would do likewise. Aaron used to be a very fine oarsman, but he has not rowed very recently and felt doubtful. Little Goliah was eager to go in the boat, so I took him. He is 10 and it is time he was learning to manage a boat.

When we got to the mouth of the Waccamaw River it was very rough and Aaron wanted to turn round, but I would not appear to understand his desire. I exclaimed : —

"Now, Aaron, you see why I wanted you to row me. I knew there would be half a gale blowing out here, and I would not have been willing to cross with any but a first-class boatman."

"Miss, you t'ink we kin mek 'em? Dem wave is putty tampsious! You see de win' is ded gen de tide, en we bleege to cross right een de teef uf de win' !"

"Yes, but the tiller ropes are strong, and I can keep her head on the waves and watch my chance to quarter over. The boat is stanch, and I promise you I can keep her out of the trough. You know the river well; tell me the best place to cross, and let us go," for all this time we were dancing about in the mouth of the creek, where it would have been easy to turn — when once we got into the rough water we could not — and I feared that Aaron's caution might prevail.

The river is about a mile wide at that point, and it certainly did look angry. Poor little Goliah was so frightened at the swirling waves that I told him to sit down in the bottom of the boat, which he did, and covered his eyes with his hands so as not to see the raging water. He just shivered when the spray dashed over him. It was a strenuous half hour, but we made it, and when we got into the canal mouth on the other side Aaron laughed aloud with pride and delight; he rested on his oars, and taking out his bandanna, mopped his face streaming with sweat and chortled with joy.

"My Lawd! 'Tis a good t'ing ter travel wid a pusson w'at hab a strong heart. Miss Pashuns, you bring me over dat ribber! I didn't trust fer cum, but you bring me."

"I know you are glad, Aaron."

"Too glad, E mek me feel too good, I got back me y'uth."

I got out on the wharf, very tremulous in my arms from the effort, but as happy over it as Aaron. I told him he must wait until sunset to return, for the wind almost always falls then. I found J. G. waiting with the surrey and was so very glad I had persisted in coming, for he said he would have waited until night for me.

Met with a delightfully cordial welcome and a dinner of delicious sailors' choice, fresh from the sea.

THE RECTORY-ON-THE-SEA, July 8.

It is too delightful here! Words cannot express how much I enjoy this beloved sea, the invigorating breeze, and the smell of the ocean! I did enjoy my night's rest so much with the glorious boom of the waves breaking on the beach, which I have not heard for so long.

The family are charming, and go on with their various occupations, and I just sit on the piazza pretending to embroider a shirt-waist, but in reality just drinking in the beauty and goodness of that "great first cause, least understood," as Pope expressed it, whose purposes we read awry, whose mercies we so often mistake for punishments, whose wisdom we so often doubt, and whose hourly call for our hearts we refuse, and still he lavishes his beauty and goodness upon us!

Sunday, July 10.

This morning coming from the dear little chapel of All Saints on the sea-shore, where we had service, I met Mr. L.,

The roof of the house on Pawleys Island — from the sand-hills.

and had the offer of a magnificent St. Bernard dog. I certainly am fortunate about dogs. My only fear is that he and my fine red setter may fight, for they say he is hard on dogs, though very mild to human beings. He is a beauty and would be a great possession to me. I feel quite sure he would not fight MacDuff, my terrier, for he has the gift of winning love from all, man and beast. Don, the setter, who is jealous of everything else, has never been jealous of him.

In the afternoon I drove with Mr. G. up to the negro

mission at Brook Green, nine miles. It is a pretty, churchly little chapel. He asked me to play the very nice organ. The vested choir of colored girls had already put up the hymns "I Need Thee Every Hour," "Crown Him With Many Crowns," and "Sun of My Soul." They sang very well, showing Mrs. W.'s careful training, and the service was very pleasant.

Visited the recluse for a few moments — a striking and interesting figure — then the homeward drive through the thick woods. Altogether it has been a perfect day.

July 11.

My time is up on this delightful beach and I started home, driven as far as the river by my kind host and hostess. Found the Whiting with Aaron and Goliah waiting for me. It was very, very hot. I steered at first, but could not hold up my umbrella and steer, and as Aaron said he had taught Goliah on the way home I changed my seat and turned over the ropes to him. He did beautifully. The river was like glass, a great contrast to the trip over, but the creek called Squirrel Creek through which we go is so winding, with such sharp turns, that I did not suppose Goliah could get us through without striking the shore once, but he did, and I was much pleased.

After the hour and a half row I looked over the corn, cotton, etc., at Cherokee and then drove rapidly to Peaceville, I was so hot and tired. As I got out of the buckboard I saw my precious little dog lying under a tree very ill.

As I called his name he tried to jump up, but could not stand and fell over on his side. I was terribly upset. I had a tub brought and poured bucket after bucket of cool water fresh from the well over him, then rubbed him dry and gave him three tablespoonfuls of olive oil. Then before going to bed six hours later I gave him a dose of castor oil in hot milk. I feel very little hope of his recovery and am very sad.

July 12.

Got up at 5 and went out at once to see after MacDuff. He was not in his bed nor could I see him anywhere in the yard. I feared he had wandered off to die — that is the dog's instinct, the call of the wild, I suppose, to go off into the woods and unseen give up its last breath. I sent Jim to search the whole enclosure, which is large, and a creek runs at the northern side. I stood a while silent by the well and then lifted up my voice and called, "MacDuff, MacDuff!" when around the piazza and down the front steps clattered the little fellow, frisking and jumping, apparently perfectly well. I am thankful; I would have missed my little companion sadly.

This afternoon Goliah came to me looking very solemn and asked to go home for two days. When I asked why he wanted to go he said his little brother, Tillman, was dead and he wanted to go to the "settin' up" and the funeral. Poor little strangely named fellow; he never was well. The same disease that carried off his mother and brothers gripped him early. I ought not to let Goliah go, but it would break his heart not to, and so I let him go. After all, poor little Tillman is safe, and this smart, good little Goliah, whom I teach and train as well as I can, is already showing that he will soon break away from my authority and he may grow up a bad man after all, while poor little Tillman is safe from evil influences. There are many things worse than death.

PEACEVILLE, July 25.

The field Loppy has ploughed is a sight to make one weep. Great boulders of earth much bigger than his head lie around as though tossed there by some giant playing ball, and the earth being dry and caked the harrowing does not have much effect. Bad as Gibbie is, this is worse. I am sending him all the nourishing food I can find to get him on his feet again.

No chance of a stand of peas with such work. The only cheering spark is little Laycock, who every other day with great flourish of trumpets deposits a tiny little egg in the geranium lined nest.

This evening I had all the children in the village to dance here for an hour. I told them I would be happy to play for them from 9 to 10 o'clock every Friday evening — not a party, because there are no refreshments, only a dancing class. They seemed greatly to enjoy themselves.

There are not more than fifteen all together, and L. came to help me direct the dancing. I am so fond of young people that it is a pleasure to me, and they do have a very dull time, especially those who have come home from school. I meant to make a tennis court in the yard, but I attempted to have the grass improved. It was moderately smooth before, but I ordered it very carefully ploughed while I was away and fresh grass seed planted. Gibbie was the person to do it, and it is now like the billows of the sea, so that a tennis court is impossible.

The mail brought me to-day a most interesting looking parcel with forty-two cents postage on it. I opened it slowly and with much satisfaction. Is there anything more delightful than an unknown quantity? When I opened the box, about six inches square by two high, out came a white canvas bucket with stout rope handle, capable of holding a peck.

I examined it with great interest and wonder as to its purpose. A water bucket, I concluded, so I called Lizette and had her take it to the ever flowing well and fill it. She brought it back held at a respectful distance, for the water dripped out very steadily though not fast. Then I decided it was for me to pick fruit and vegetables in. I could only see by the postmark that it came from Maine. I am quite charmed with its lightness. No basket is half so light to carry.

To-day Chloe is walking about the yard a little, which is a joyful sight to me. She at last got a chance to tell me her amusing story. One day while I was away, Gibbie came to her looking most mysterious.

"Cook," he said, "I got somepin' fo' tell yo' 'll 'stonish yo'. I study 'pon um till I confuse een my min'. I dunno ef I kin tell yo' straight, but anyhow I'll try. Yisterday my wife en all de 'omans on de place, gone fer chop cotton to Mr. O., en dem bin' a wuk en him wife run out en 'e say: 'So Miss Pennington hab fer giv' up plant cotton altogeder,

"En de 'omans mek answer en say: 'No, ma'am; we neber steal none.'"

una steal um so bad! En de 'omans mek answer en say: 'No, ma'am; we neber steal none.' Den de lady gon' een de house en bring out a newspaper en read out de newspaper, en please God, an' Chloe 'e read out o' dat newspaper eberyt'ing w'at happen on dis plantashun!

"De 'omans ben dat struck dem was same like a dumb pusson — dey was all de gwine-on 'bout de cotton-fiel'. De paper tell how dem gon' een de fiel', soon ne mornin' en pick de cotton, en w'en de moon shine how dem pick de cotton, en how dem mek pilla en bolsta, en at las' mattrus out de cotton, en ebery free-male been struck. 'Kase dem know 'twas all de truf. W'en my wife cum home en tell me I had to mek him tell me ober t'ree time befo' I git de onderstandin', en I ain't dun study yet how come dat kin all bin een dat paper."

Chloe told it with much more dramatic force than I have. He went on : —

"De whole plantashun stir up. Some say dem g'wine 'way, say dis is a witchcraf' place. Kyant onderstand how all dem sekrit kin git een de newspaper. De only t'ing all de name different. I kyant remember wha dem call Uncle Billy."

Chloe asked if there was anything about him.

"Yes, say how him seem like him couldn't ketch up wid de people, say him do all he kin but him cudn't seem to manige dem."

Two days after that, Chloe says, my good little Georgie came to her in great wrath. She had been at the reading and repeated the whole story to Chloe with small variations and a good deal more minutely. Her indignation was so great that Chloe tried to pacify her, but she would not be pacified.

"What hurt me is that I ain't got a single pilla nor nothin' f'm de cotton," she said. "I got my two lone fedder pilla I had w'en I married, en ebrybody else got dere house chock full, en yet de disgrace fall on me same like on dem."

Then she went on to recount the fine bedding all the others had. At last Chloe said : "Well, Georgie, no one kyant help it; ain't yo' know dem ben a-tek cotton fum de fiel' all de time?"

"To be sure I know, yo' cudn't help know."

"Did yo' eber tell Uncle Bonaparte 'bout dat?"

"No, I neber tell nobody."

"Well, den, you kyant say not'ing, en ef yo' only bin tell him onct, yo' name would be clear; but now yo' kyant git mad 'bout dat, kase yo' neber clear yo'self."

It was a new view to little Georgie, and helped somewhat to pacify her.

When Chloe left me I thought over it a long time, but concluded it was best to take no notice of it in any way. The hands had all been a little on their dignity: but I was pleased at that, because they did better work to sustain the dignity, and that is all I want.

July 26.

A perfectly delightful temperature, so cool that I had to put on my white flannel suit, made from my own wool, which is very warm. Old Daddy Ancrum came and I was so glad to see the old man; after giving him a good breakfast, got him to work out the peanuts, which he did beautifully. He must be nearly ninety and yet does such beautiful work and takes such pride in it.

He says Bonaparte is a child to him, and Bonaparte was born in 1833. I wish the old man's farm was nearer. It is quite a large tract and he has given a part to his son, Kilpatrick, who is a carpenter. If I could get Ancrum to superintend the hoe work here it would make all the difference in the world in the results. But he is greatly interested in his own farm and only comes now and then when he wants something.

My rice is beautiful, contrary to all expectations. It is upland rice and has stood the drought better than any of the other crops. Jean and Florinda have worked it perfectly clean; there is not a spear of grass and it is a rich dark green and growing apace.

I have Goliah at last in whole clothes. I had a very stout piece of sky blue denim, and his first trousers were made of that, and with a blue and white shirt he is quite startling. Then he has two white suits. I choose white because I can see when they are clean, which I could not do if they were dark colored. He is very proud and has redoubled his activity.

He is so small that he has to have a box to stand on to harness the horse, and even with that he cannot get the head-stall on without help. He is very persistent and very gentle with "Root," as he calls her, and I admire the graceful way in which Ruth has yielded to him. She really tries to help him in every way and stands stock-still while he labors with the fastenings of the collar and hames. Goliah has seen a good deal of life and he feels that just now the lines have fallen in pleasant places for him, and he does his little level best all the time.

On Sundays I take him to the church gate in his sky blue suit to carry my music books for me. The first time he went he had a little wistful look, so I said, "Would you like to go to church, Goliah?" "Yes, ma'am," he replied. So I took him in and showed him the pew reserved for his color and told him to watch when people knelt and stood and sat, and to do the same. As I sat in the choir at the other end of the church I had to exercise my faith in his discretion. When I heard him say his little catechism that evening he told me he "'joyed the chutch mutch. Befo' I never cud stand to go to chutch, but I like dis, en I want you, please, ma'am, to le' me go next Sunday." Of course I was very pleased, and ever since he has gone to church and I am told by a most particular member whom I asked to give an eye to him that he behaves perfectly.

I was so pleased with this that it was a shock to me to find that Chloe disapproved intensely of it. When I asked her to leave the dinner for Patty to cook the last Sunday our minister was here she said no, she did not feel like going.

I urged her to go, when to my amazement she said, "No, ma'am! You t'ink I'd go en set down by dat chile een chutch? No, ma'am, if I neber go to chutch I wouldn't set down by Goliah!" I retreated before the unknown; you may live near these people all their lives and never understand them. Goliah is preternaturally clean, for I have to take him about in the buggy with me, and that is why I have him wear white entirely; there is no concealing dirt on a white suit. So that the scorn of sitting beside him comes from something different and incomprehensible to me.

Poor Jim is terribly discouraged. The corn is being stolen daily.

After these rains the track of the thief is plainly to be seen, a very big, bare foot. Jim called me to see it and I took a little cane and measured the track and when I came home took my tape measure and found it was fully thirteen inches long. A smaller foot is also visible.

Lizette tells Chloe how grand a time every one in the street has at night with big pots of corn boiling on the fire and even the babies eat it. What hope is there of ever making, or rather getting, a crop of anything? They are as natural and unrestrained in getting at what they want to eat as ants, and just as hard to frustrate and control.

Sunday.

This morning Goliah said he wanted to get off early as Jean was to be baptized.

"Where?" I asked eagerly.

"Rite een de ribber, up to Belside."

"Oh," I said, "wait a minute; I must send her some things," and upstairs I flew and turned my bureau drawers topsyturvy and found a complete outfit, a white lawn skirt which is one of my prime favorites, having a deep flounce around it, a white lawn shirt-waist, collar, and belt.

Poor, forlorn Jean, whose life I saved three years ago when

she seemed a certain victim to tuberculosis — and poor thing, I sometimes wonder if I did her a kindness, so undisciplined and unfaithful to every duty does she seem. And now to hear of her being about to step into the river and wash away her sins!

I was greatly excited, and with trembling hands, for fear I would not get them to her in time, I put up the parcel and sent Goliah off at a full run.

July 28.

Another perfect morning. I read last evening an article on efficiency which dwelt upon the necessity of relaxing, not pushing on, nerves and muscles taut and strained all the time. That is my snare. I was much impressed and determined to relax to-day and take a complete rest at noon. I carried out my intention and relaxed, with the result I never braced up again! Never was able to do a thing for the rest of the day.

July 29.

Had a very trying day — not money enough to pay off the hands in full, and that always demoralizes me. I went down in the field to examine the work. I always walk now, since reading an account of a visit to the work on the Panama Canal, the writer having been nearly killed by the length and rapidity of the walk, Col. Goethals saying, "If one wants to keep well in this climate he must walk." Since then I make it a point to walk a mile every day.

My own want of efficiency worries me. To-day again I relaxed and rested, and I know it was a mistake and will not try it again — some people have to stay braced.

Lizette, who is about fourteen, went last night to a "settin' up" three miles from here. A woman had died whom she did not know at all, had never seen in life. In the midst of the singing of "speretuals" and shouting two small boys got into a fight, their parents joined in, and in a few moments the "speretuals" and shouting were turned into cursing. Poor things, poor things! Lizette was so worn out that when I came down to breakfast I found her stretched out on the pantry dresser fast asleep.

After breakfast Chloe came in and told me she was freezing cold and could not get warm. I immediately went out to the kitchen and made a cup of hot ginger tea, which I forced her to drink. I tried to get her to go to bed, but in vain; she said if she once went to bed she knew she would never get up again, and this melancholy view I did not combat. I just said: "Then perhaps you had better stay up."

I made Jim cook as Chloe was too ill to do anything, though she would not leave the kitchen until I had her big rocker brought and put under an oak tree just in front of the kitchen and insisted on her sitting there. Goliah was made to put on his white apron and wait, which made him very proud.

God forgive me; but it does seem so hopeless when the elements are banded together against one!

I must remember this is the time to show faith and courage.

Sunday, July 30.

The blessed day of rest. I wrote that this morning. It has been a blessed day, but not one of rest exactly.

I had early in the week a letter from C. saying he would bring the dean up this afternoon to have service at St. Peter's-in-the-Woods, about nine miles from here, asking me to meet them there and saying they would come home with me and spend the night. I think I did too much Thursday, driving. Anyway I was very nervous.

I let Jim go down to Gregory Friday and spend the night with his family, so that I could have him here to-day to drive me. I fixed all the lamps and finished my household work, for this tall Lizette cannot be trusted to do anything. Then at 10:30 got into the wagon behind Ruth and Marietta to go to church in Peaceville.

I had been invited to dine by Mr. F. and M. had a delicious dinner. Then I took them with me out to St. Peter's-in-the-Woods. There was a very small and pathetic looking congregation. The notice had been short. Mr. S., who had promised to give it, had not been very successful. These people do not go to any post-office or have any mail, so any notice to reach them has to be sent by hand to a few in time to have the word passed round.

When the dean drove up with C. I saw him look around with wonder, first at the very forlorn looking congregation talking together in groups, and then at the very plain little board building which is the church, standing in a group of trees on the edge of a swamp. I realized at once that the eloquent divine had never come upon just such a church and just such a congregation and that for the moment he was taken aback.

After a while the service began. The dean with his fine voice and in his handsome vestments seemed quite too big and imposing for the little chancel with its bare pine table and reading stand. The little baby organ which was given to the chapel years ago has long been dumb, so I had to raise the hymns. The dean helped much with the singing and read beautifully.

When the time came for the sermon he read the miracle of the loaves and fishes and then in a low, quiet voice talked. What he said was very beautiful and very simple. With that hungry multitude and nothing but one boy's individual store, our blessed Saviour might have made a great and wonderful spectacle and by His word created thousands of loaves and thousands of fishes and caused excitement and amazement; but He simply asked the question, "How many loaves have ye?" told His disciples to make the multitude sit down and to divide out what they had, and lo! they had enough and to spare.

Then he pointed the lesson to us. Do not wait for great things, do not long for great powers, for great opportunities; use the little you have in faith and God will make it cover the need; use your little strength; use your little talent; use your little store of whatever kind, and it will suffice. I cannot give any idea of the effect, but I must write down what l can so as not to forget it myself.

When I went out of church poor Betty C., whom I have known from her girlhood and who has always looked old and weary, her capacities always having been below her needs, said in her very slow, drawling voice: "Miss Patience, is this here preacher comin' here ag'in?"

"Yes, Betty," I answered. "The dean says that whenever he can spare an afternoon from his church in Gregory he will come."

"Well, Miss Patience, I'm mighty glad to hear it. Seems like I'd walk any distance to listen to him."

"Well, Betty, you tell him that; it will please him."

Whether Betty ever made up her mind to such an effort as to tell the dean I never knew. She is a woman of 46, tall, thin, bent, yellow, the mother of seven children and one grandchild. Her husband is the owner of much land and quite a stock of cattle, and plants a good farm. Her life has been one long effort to keep up with her duties, for she has faithfully tried in a feeble, helpless way to do her duty. That the sermon should have reached her heart and helped her was a wonderful tribute.

These pineland white people have a strange pathos about them, a wistful, helpless look like some spirit that would fly, would soar, but is bound securely to the earth. My, but they are pitiless to the one who falls from their standard of morals! I asked several about poor Mrs. Lewis. The answer was always with averted eyes, "I ain't heerd nothin' about her for the longest." I tried one after another, but always the same answer.

The Lewis family live but two miles from the church, just on the road, and many of them pass the hut in coming to church, so there must be something very wrong. If Louise, who teaches the Sunday-school, had been there I could have found out what was the matter, but her last baby was too young for her to come out, and it was too late for me to go to her home.

The drive home was delightful. I got home about six and was able to have supper all ready by the time C. and the dean got here. We had a charming evening and I feel greatly refreshed mentally and spiritually in spite of bodily fatigue.

July 31.

C. and the dean got off, to my great regret, about half past nine. It is my dear C.'s birthday and Chloe made a nice sponge-cake in honor of it.

After they left Chloe began to pour out a sad tale about Goliah. I had forgotten to give notice that I would not be here for the Sunday-school in the afternoon, and the children had arrived as usual at 4 o'clock and Goliah had conducted them down to the garden and she hearing great sounds of mirth and revelry went down and found them all with as many peaches as they could possibly carry.

Of course she was very indignant and scolded them, Goliah specially, whereupon Goliah's sister Catty, who is well named, being of a feline nature in the worst sense of the word, had broken out and "cussed" her outrageously. Altogether Chloe seemed anxious to impress upon me that my efforts to teach them were quite thrown away and that it was a constant danger to have that "gang o' little niggers" coming about on Sunday afternoon when she was away usually.

She said she did not take the peaches from them, as it was Sunday.

I told her I was glad she did not. It is very hard on Chloe to see the peaches which she has watched with such pride and picked so carefully so that I may have a few every day as they ripen, taken off by the bushel in that way, and I feel for her. The one faithful person does have a hard time.

All these years I never had any fruit, but this summer I have had since the last of June a watermelon every day for my lunch and peaches and cream for breakfast or dinner, and both Chloe and I have rejoiced in it. Besides she has made several jars of peach preserves and had hoped to make several more.

I had to console Chloe as best I could and promise to be very severe on Goliah. It is well that I had such a spiritual uplift yesterday, for things seem specially sordid to-day. I wanted to do some writing, but the little vexations were too numerous and engrossing. Woe is me not to be stronger, to let myself be made useless by these gnat stings.

I went down to the field and found Rosetta and Anna and Becky doing good work, also old Florinda and Jean. Then I came back and did some necessary mending, and by afternoon quite late I went down to my table by the river with the dogs and got back my serenity and ended the day by working round the tomato plants.

Before he went home at 6 I called up Goliah and gave him a talk, told him how hurt I was that the children whom I was trying so hard to teach the beauty and worth of honesty should behave so. Then most unexpectedly Goliah took all the blame and said: —

"Need not to blame de chillun, Miss Pashuns; not one bin een de gaa'den but me. I gone over en I pick de peech en I give em to de chillun. Dem all stan' outside de fence en I give evry one as mutch peech as him cud tote."

This astonishing truth telling raised my spirits greatly; if Goliah had broken one commandment he was coming out nobly in telling the truth and not bearing "false witness against his neighbor." So I told him how glad I was to hear

"Dem all stan' outside de fence."

that he alone had been guilty, but he must never be so liberal with other people's things again.

Altogether I am going to bed in a very happy frame of mind. Chloe came in after tea to talk and I tried to cheer her by telling her of Goliah's confession, but it seemed only to add fuel to fire that he should have the impudence to tell me to my face that he had taken all those peaches, stripped the tree, etc. I am glad I cannot understand her point of view. You cannot blame a person for being color blind or near-sighted; you are just sorry for them and thankful that you have the joy of seeing the distant clouds in all their gorgeous sunset hues.

A week ago I had a letter from Rab, whose four years under the worthy Jenkins have still three months to run, begging me to bring him home. It is a funny little letter. After the usual politeness which a darky letter never leaves out, he says: —

"I t'ank you mutch, Miss Pashuns, fur all you done fur me, but now the time is come fur me to do fur you, en I kin help you if you'll only send fur me to come home. I kin milk, an' I kin plough, an' I kin drive purty well. If you don't send fur me please come to see me right soon."

I was quite touched and felt like sending for him at once, though the time I brought him home after typhoid fever was most unsuccessful, for he shot himself and as soon as he was out of the doctor's hands I sent him back. Now there is a difficulty as to where he would sleep if he came back. I approached Chloe as to fixing up the little room off of her room for him, but she was so upset at the idea that I gave it up. I know it would be folly to put him alone in a house outside; he would simply have every vagabond in the country to sleep with him or spend his time running over the country at night, which could only lead to trouble. I am disappointed in Rab's scholarship. The handwriting is passable, but everything else is pretty bad.

August 1.

After breakfast was going to the field when Chloe came in terribly excited and said she was going into the field to beat Jean, who had told a lie upon her. When Chloe's blood is up there is no stopping her; I really was frightened, for I did not know what would happen.

I went to the barn-yard to give out the feed when Chloe returned, dragging Jean by the arm. I sent for Lizette and Goliah, heard the whole story, and held court. A complete lie Jean had told, saying Chloe had met her at the pump yesterday and told her that Lizette and Louisa said Jean's clothes never looked well on her because they were all stolen. By this time all the hands had assembled and there was quite an audience. Chloe, still holding Jean tightly by the arm, asked: —

"Did I speak to you at all yesterday, Jean?"

"No, mam," said Jean.

Chloe said, "Lizette, you yere dat?" And so on it went, Jean confessing that it was also false what she had said, that wretched Goliah had said he would cut Chloe open with a knife.

After Chloe had said her say, I said a few words in judgment: that they could be indicted but I would only require that they ask Chloe to forgive them for their outrageous conduct. To my great surprise Jean stepped forward and said: —

"I beg your paadon, An' Chloe," extending her hand, which Chloe took and with a grand air said: "I fergiv' you, Jean."

Thankful to have the court so peacefully adjourned I came back to the house and spent the morning cutting out Chloe's "fine black" frock and an extra waist. I always offer to cut out for her, but it ends in my stitching it all up on the machine.

It has been another perfect day and night. How wonderfully good of the Creator to make this world so beautiful.

August 4.

Read till twelve last night; I felt the need of relaxation and diversion. Finished "Routledge Rides Alone," which I have enjoyed immensely, though there is too much war for me.

The working of the potatoes is almost finished. Jim is cutting tops of the oldest corn. I always have a fight over the cutting of the tops, but I insist on doing it because it makes nearly twice as much forage.

Nelly Thompson paid me a long visit. I had just washed my hair and was busy sewing while it dried so I asked her up into my room, which pleased her much. She is the widow of our faithful and devoted servant Nelson, and I always like to do her honor in a small way, though she is not at all made of the same clay as her husband.

When she was going I presented her with an embroidered black scarf of mamma's about two yards long and three-quarters wide and she was very happy. From the dates she gave me she must be 78. She is Jim's mother-in-law.

The first of this month I sent a notice to the five young men negroes, who rent houses on my place, that they must pay up their rent. The agreement was that they should pay $1 every month for the house and three or four acres of garden and field. No one has paid, and I must have the money or their work.

I thought writing a formal letter might do what speech had not accomplished, but as yet there has been no result. I want to gather my fodder, and need all the hands I can get, to do it before the weather changes.

August 14.

Went to Casa Bianca to-day. The peas are up nicely there, though the ground looks very rough. At Cherokee the men gave me one day's work on their rent as a great concession and I got in the fodder.

Poor little Laycock has made up her mind to sit on her beautiful little nest of eggs and no power can stop her. There she sits through the broiling heat of the day, and there is no hope, for a pullet's first eggs never hatch. I have tried every method known to stop her, except ducking her in water; that I would not do.

That instinct of fulfilling one's destiny and duty is very wonderful. The heat has been fierce, and the box is in the full sunshine, the scarlet geraniums in full bloom, and at first her little comb was as red as they, but gradually it has paled till it is a dull gray now. There she sits gasping with her poor little mouth wide open. I put water by her, but she does not avail herself of it.

Twice a day she flies down and goes to the well, and is fed, making a tremendous ado among the other fowls, and then back again to her post. It is a thing to excite the most enthusiastic admiration, that adherence to the post of duty — duty for its own sake. I am going to have the box taken out under a tree, where at least she will not suffer so.

August 22.

A negro man and his wife asked to see me early this morning, whom I did not know, though they were descendants, both of them, of our own people, and I wondered what they could want. My joy was great when I found they wanted one of my heifers, and had brought the money. The man was bargaining with me trying to get it for $10 when the woman said : —

"Miss, I may's well tell you I wants one o' your breed o' cow, en I got de $15 in me pocket fo' pay fo' em."

So I sold her my beautiful Dodo, and I promised to keep her in my pasture until their fields were open. It is a mercy to me to have this unexpected sale.

Last week I sold Peacock with her picture calf. I nearly wept, but yet was glad to get an offer for her. She was a pale gray, with eyes all over like the feathers of a peacock and a splendid milker.

I don't allow myself to put down anything about the heat; after seeing little Laycock's unmurmuring endurance in her maternal zeal, I feel it is unworthy to dwell upon or even mention the subject, but it has been terrific.

Yesterday, as I drove down, at one place where the pines are thick a splendid wild turkey rose just in front of us and soared away, greatly to Ruth's alarm. To-day very near the same place we came on a number of them. There must be a growth just there of the berries on which they feed.

I had the wagon and men at Hasty Point landing at daylight waiting for the steamboat, which was due there at that hour. It did not come until twelve, but the church organ was on board and put at once into the wagon and brought out here, where Miss Penelope and myself superintended the unpacking and had it put into the church. Just as this was done there came a downpour of rain. I am so rejoiced that the organ has been restored to the church and is now in perfect order. This great blessing we owe to a generous friend at a distance, who this spring sent the money to pay for the repairs and freight.

Sunday.

A perfect morning. Oh, the joy of this blessed day of rest and peace! That the Almighty One, who needed no rest, whose powers are infinite, should have ordained this seventh day of repose and cessation from toil, seems too wonderful. As I sat at breakfast (a plump little summer duck) and looked out into the depths of foliage, all shades from the solemn, steady green of the great live oaks through the wild cherry's shining leaves, the Pride of India's diaphanous fronds, the walnut's dull, yellowish, palmlike branches down to the vivid apple green of the grass — all so perfect, so full of beauty and delight for the eye of man — on this His day, here in my isolation the love and mercy of God and the joy of His great gift of life intoxicate me. I feel as David must have felt when he wrote some of those glorious shouts of joy and praise. I long to give expression to my overflowing gratitude.

Monday.

A dreary day of rain, which I found it hard to get through. This is a sad season to me. I do not believe in keeping anniversaries, but they hold one in spite of every effort. Even when there is much of interest going on around, there is deep down within the heart that nag, nag, nag of memory, like the toll of a bell, every day, every hour, every moment of the agony, thirty years gone by. The maddening "Why was not this done?" "If only that had been done!" and so for the time one forgets God and His everlasting arms and centres the mind on poor human agencies and possibilities. One cannot read, one cannot sew, one cannot pray.

CHAPTER XII

August 27.

DEAR old Daddy Ancrum came dressed in his Sunday best to tell me all he could remember of his past life. I had asked him some time ago to come some day when he felt quite well — but I was quite touched at his dressing in his very best for the occasion. It was most interesting to me and I wrote it all down. According to the dates he gave me he is 91 years old — with all his faculties and in good health.

Next Sunday there is to be a "funeral sarmint," preached for Chloe's aunt, a person of distinction in the colored world, and Chloe and Patty both want to go. I will keep Goliah, so as to have some one in the yard.

As we drove to-day I asked him if he could cook rice; that if he could cook I might have him do so Sunday. He said he could, but as he would answer that to any question asked as to his powers I asked him to tell me how he did it. He began: —

"Fust t'ing yo' roll up yo' sleeve es high as yo' kin, en yo' tak soap en yo' wash yo' han' clean. Den you wash yo' pot clean, fill um wid col' wata en put on de fia. Now w'ile yo' wata de bile, yo' put yo' rice een a piggin en yo' wash um well, den when yo' dun put salt een yo' pot, en 'e bile high, yo' put yo' rice een, en le' um bile till 'e swell, den yo' pour off de wata, en put yo' pot back o' de stove, for steam."

I was so impressed with the opening sentences that I determined at once to let him cook my Sunday dinner instead of eating it cold, but when I told Chloe she was filled with indignation.

"Miss Pashuns, if I neber eat rice again I won't eat rice Goliah cook! But den I'se bery scornful!"

Fanning and pounding rice for household use.

August 29.

Chloe and Patty went to the funeral "sarmint," and it was grand. The eulogies of the departed were satisfactory to all. They left in the buckboard at 10 o'clock and returned at dusk, the church being six miles away. When I asked what the preacher said about Elsa, Chloe answered: —

"'E set um high, but eberybody groan an' say amen to ebery wud. Fust t'ing 'e say she wus a fair'oman; what 'e had to say 'e say to yo' face. She wusn't tale bearer, she wusn't 'struction maker. She wus a stewardness of de chutch en always fait'ful. She house wus a place fo' de preecher en de elda' to fin' a home w'en eber dey kum. En de feebla' en de olda' she husband git, de mo' she was 'evoted to him; nobody neber hear um say she tyad, nite en day she nuss um; she was a wirtue to im, en a sample to de yung womens."

I could not help thinking Solomon could not say more for the woman whose value he set above rubies.

I have had a very peaceful day. I did not feel strong enough to go to church. Goliah boiled the rice beautifully, and I made my dinner of rice and milk and rested. The heat has been fierce lately and I feel wilted, but the first autumn month will soon be here.

September 1.

The papers tell of floods everywhere, but they have not yet reached us. The Pee Dee is reported higher at Cheraw than it has ever been. It takes its rise in a spring under Grandfather Mountain in North Carolina, and so a rainy season in the mountains or melting snows always give us a disastrous freshet, now that the banks have been stripped of trees, the whole of its long and winding course.

Pounding rice.

September 3.

Bonaparte sent me word last night that the water had made a great rise during the day and I had better come down early.

I could not get breakfast in time, so I took a glass of cold coffee and a piece of bread, and went.

The barn with my rice, a very large two story building, was surrounded by water about fifty feet. With the aid of a boat Bonaparte made a very swaying bridge and I went in with all the empty sacks which could be gathered, and measured and bagged the rice, removed a plank from the flooring of the barn above, and had the seventy-five bushels taken up.

There was no time to call hands, as the water was rising rapidly to the floor where the rice was piled, so Gibbie and Dab with Bonaparte did it all. We had an active and very dusty morning, but unless the foundations of the barn give way the rice is safe.

I was perfectly charmed to find I had so much, for I have been eating it and paying for my work with it and trading it for two years; it has been a perfect widow's cruse.

Coming home the clay gully was so high that the water came into the buckboard. Ruth didn't like it and pulled until she broke the harness, but we got out safely.

September 4.

All the roads we usually travel are impassable, the bridges under water or floating. The men go through a cart path which avoids the bridges, but it is a roundabout way and does not help me a bit, so I just plunge through the clay gully every day.

All the men are in despair. The entire rice crop is about four feet under water and there is very little hope of saving any. I have been so unhappy because I had not planted any rice and accused myself of supineness because I was afraid of going into debt to plant any, and now I am so filled with thanksgiving that I didn't, and feel that I was specially guided not to do it.

September 5.

Poor Gibbie is so determined not to work that he has broken the plough. I was very anxious to have land for turnips and rape prepared and Gibbie could only get out of it by breaking all three of the ploughs. Then I got him to mowing the hay, and he promptly broke the mowing machine.

The cause of these mishaps is Gibbie's sporting tastes. All the men on the place, who will not pay their rent, are over the river with guns making large bags of game of different sorts. The poor rabbits and other things having taken refuge on any knoll or stump and are easily shot as the water recedes.

I might as well give up any effort to have work done until the waters have entirely subsided. The damage from the freshet is wide-spread, but thank God! no loss of life or cattle.

September 6.

In the corn-field all day. We never gather and house corn as early as this, but the stealing is so much worse than usual that it is either now or never. I could only get the women out, so I made Goliah do the hauling. I rode Romola, and she was very disagreeable and restless.

September 7.

In the corn-field yesterday and to-day. A perfect day, and the air crisp and not too hot. Oh, the beauty of the sky and air and trees and the black-eyed Susans and goldenrod everywhere! Oh, the mercy and goodness of God in making all this beauty and showering it on us unsatisfactory, discontented, grumbling mortals!

As I sit under a tree and drink in all this beauty and wonder, I resolve never to think myself hardly used, never to long after the yellow gold which greases the wheels of the world and makes life so easy, while I have all this golden glory of beauty and sunshine, and the power to see it and enjoy it to the full.

The week's stay in the field has rewarded me. I have in the barn 550 bushels of corn and about two tons of sweet, dry hay — only the first cutting and not a drop of rain on it.

Last Wednesday Gibbie asked me to lend him my canoe. I hesitated, for it is a very nice white boat. He said: —

"Miss, ef you'll len' me I'll be keerful wid um en I'll gi'e you some bud fo' pay fo' um ebry day."

I at once consented to let him have it if he would give me a dozen birds as rent.

The next morning I went down to the plantation with the pleasant expectation of having a nice dinner; the rice birds are tiny, but delicious. When I was leaving I asked Gibbie where my birds were, he brought out three and said they had had a poor night's sport.

The next day he said there were none, as there was no dew and they could not get them when there was no dew. Friday night there was too much wind, he said.

On Monday, when I asked him, he said he didn't go out Saturday night, as it was too close to Sunday and he had to prepare for church. I thought that quite proper and only said incidentally, as it were: —

"And you did not go last night?"

"Oh, no, ma'am; not Sunday night. I wouldn't do sich a t'ing."

I thought how careful Gibbie was to observe the Fourth Commandment, as they begin their operations about 2 A.M., but I said nothing and left the garden where he was working. As I left he turned to Goliah and took out of his pocket a five-dollar bill and said: —

"Look w'at I make Saturday en Sunday night!"

Goliah told Chloe how Gibbie had showed him the money and told him that no night since he had the boat had he made less than $2. This evening I told him I could not let him have the boat any more, as he had been so unsuccessful.

He raised his voice and declared solemnly he hadn't been out but the one night and had given me three birds, which he intimated was handsome, and talked on in an injured voice. I only laughed and said I was sorry he was so unlucky; that I could not lend my boat any more.

They take lightwood torches and thresh the bushes with a long rod or switch and kill the birds, often getting two or three bushels in a night, which they sell to men waiting on the banks for 35 cents a dozen.

I have made another effort to get the men to pay their house rent now that they are making so much money so easily, but in vain. As some one said to me the other day: "I never realized the power of a lie until recently! Any one who can make a plausible lie and stick to it, seems impregnable."

It is an awful thought, for we know who is the father of lies, the Prince of Darkness. There is no shaking my faith, however, in the ultimate triumph in that never ending, to-the-death struggle between the powers of darkness and light, — that the light will conquer every stronghold of darkness until the perfect day reigns the world over.

CHAPTER XIII

September 8.

ROSE at five and read the lessons on the piazza and then churned. There is certainly a wonderful freshness and life in the early morning air, a kind of inspiration in watching the birth of a new day. I get terribly hungry, however, before I can get any breakfast. This morning a delightful waiter arrived. It had shrimps and flounders fresh from the sea and great yellow pears with one red cheek.

I did not go to the plantation, so had a day off and enjoyed it thoroughly. I have a most delightful book which I have been pining to read, but had to resist until to-day. It is the life of Alice Freeman Palmer, and no words can express the refreshment and uplift it has given me.

I wish I could give the book to every young woman in whom I am interested.

September 9.

A brilliant morning. I tried to get to church in time and succeeded. All the invalids out, which was such a comfort. Our rector gave us a very good sermon on prayer. There was a terrible mix-up in the choir in the "Gloria in Excelsis." I sang one, while Miss Penelope played another! The results were truly heartrending, which was a pity. Still, the intentions were good, and we were both so in earnest, that we could not stop, apparently. The worse the sounds were the more we persevered.

September 10.

Every effort that I have made to induce the men to pay their rent has been vain. Last evening as I was coming back from Casa Bianca late in the afternoon, feeling very discouraged, I saw Green ahead of me carrying a pair of wild ducks and a string of coots. He was going toward Peaceville and I had a moment of satisfaction, for I thought he was taking them to me to pay on his rent. So as I came up with him I said in a cheery voice: —

"My, Green, I am so glad you are bringing me those ducks and coots. I have only eaten one coot this year."

Always civil, Green answered in his softest voice: "No, ma'am, dese don't b'longs to me; dey b'longs to dat gent'man ahaid," pointing to a negro man who was walking about five hundred yards ahead.

I could say nothing. I knew it was not so. I knew they were Green's, shot on my place, and if he had given them to me it would have reduced his debt from $9 to $7.50, and though I specially needed money I was willing to take anything to help him make a start.

September 11.

Yesterday Cable came to see me. I have been sending after him for some time, but couldn't catch sight of him. After a few polite inquiries as to health he said he heard I wanted to see him.

"Yes," I said, "I have sent for you many times, I want you to cut some wood for me."

"On what 'rangment, Miss?"

"For half," I answered, "and I will furnish the flat without charge."

He said he could not cut wood for that — it was too little, $1.25 a cord did not pay him for his time. After some little talk I found that it was a waste of time to try to get the wood cut, so I said: —

"You have not done a stroke of work for me since April 9, when you helped fill the boiler; you have your house rent free, all of the fire wood you wish to use, and two acres of fine land around your house on which you have a very good crop of corn; and yet you have never been willing during these four months to do anything for me. You owe $9 for the steer you got and begged me to give you time to pay. You have had it eighteen months and made money by its work but never have offered one quarter of a dollar on it.

"Then in January you came in great distress about your wheel, said you had given it to a man to mend and that he charged $2 and that you were not willing to pay it when you had the money, and now that he had sent you word that he was going to sell the wheel unless you paid at once, and begged me to lend you the $2 and you would return it soon or at any rate work it out. But you have never paid a cent of it. In March when I paid you $7.85 for cutting wood and reminded you of your debt, asking you to pay something on it, you pleaded with me not to take anything then, you had a particular use for the money, and believing you, I consented. You say you have been getting one dollar a day where you have been working, and surely you could have made an effort to pay something on your debt.

" Now I want you to work for me; as you will not cut wood, go down to the plantation at once and tell Bonaparte I have sent you to drive the mowing machine. Jim will not be there until one or two o'clock and I want the hay cut this morning. Cut one bed, it will not take long, and then help Jim haul in what he has raked up. Go as quickly as you can, for I must get as much done as possible, the clouds are gathering."

He acquiesced at once and I told him I wanted him to run the mowing machine all this week; it was then about ten o'clock. In the afternoon I drove down and found Cable had just turned up at four o'clock; when I asked him about it he said, "Yes, ma'am, I won't tell no story about it. I did stop on the way," and that was all. Of course the hay was not cut and only two loads were hauled in, and last night there came a tremendous rain wetting it all so that all windrows had to be opened and spread out again, and to-day Cable did not come to work, and I have no possible redress or power over him. Of course every one will say I was foolish to trust him in the first instance; but I am made that way and I cannot unmake myself; if any one living near me appeals to me in distress, if I have the money at hand, I will lend to them, and time and time again I am deceived and disappointed. I said to him: —

"Neither of your grandfathers would have acted as you have done. Daddy James was an honest man and never tried to shirk a debt, but you, though you are free and have schooling and all the help that a good, industrious wife can give in your life, are not ashamed to act so."

I am more hurt than I can express by his not coming back to work. Jim has worked tremendously to try and save the hay, and now all that is standing should be cut, or it will be hard and worthless. While there is so much down that has to be dried and handled, another man is absolutely necessary.

September 12.

About 9 o'clock Chloe came in great distress to say she had just heard her Uncle Mose was dead. Chloe was greatly upset at the news.

September 13.

A real autumn morning. The first let-up to the heat. I thought in the night we were going to have another storm. I prayed hard against it, for my pea-vine hay would be ruined. This morning the east wind is high but it looks brighter.

Chloe's getting off to old Mose's funeral occupied the whole morning to the exclusion of everything else. Goliah drove her in the buckboard with Ruth.

I have the cotton which has been picked spread out on the

piazza to dry thoroughly. After Chloe got off I sat on a stool beside it and picked out the cotton, which was greatly damaged by the wet. The cotton was just ready to open when the storm came. It was arrested and kept in the close little case soaking wet, and then nature was busy and little sprouts came and went on growing as though they were in their proper element.

I took out a little pinkish tight wad and opening it carefully there, folded up in the middle, I found a little green leaf. Of course these would injure the quality of the cotton, yet it was impossible to have them kept separate in the picking, so I pick them out when it is spread out in the sun.

While I was doing this Jim came to get orders and I had him pick a while. I commented on Goliah's delight and excitement over a death and funeral. He had been so sick and miserable yesterday that he went home about 4 o'clock, but being sent to tell Chloe the news he came to the yard and then took her out to the street and escorted her back here again about half past ten. He was eager to go to the "settin' up," six miles up the road, but could find no one willing to walk up there with him. I said: —

"I do not think there is anything in which the races differ more entirely than in their attitude to death. No white child wishes to have anything to do with death; they fly from the signs and tokens of it, whereas the children of the African race seem to be attracted by it."

Jim answered: "What you say must be true, Miss Pennington, for nothin' gives me more pleasure than to handle the dead. I just delights in it an' I have great luck in it, too; scarcely a person dies but what I have the privilege of jumping down in the grave and receiving the body an' makin' it comfortable in there."

I was quite startled. He went on: —

"W'en I cum frum town an' fin' that they'd buried Georgie

2 c

without my bein' there to handle her I was that disappointed I couldn't scarcely stand it."

He had gone to town before Georgie died one afternoon and she was buried the next before he came back. It is curious to see racial peculiarities continue from generation to generation. There is no repulsion to or fear of death among negroes as long as the clay is visible, but as soon as the funeral is over and the grave is left, then terror begins. Jim himself does not like to walk down the front avenue alone at night, because it is so near the beautiful spot where those of his race who have died here, have been buried for over a hundred years. And still they come. It does not matter if they have died elsewhere if they are prosperous, and even if it is a mighty effort they beg to be brought "home" and laid by their people. As my father owned 600 when the war ended, it makes a number of funerals, for all the descendants of those want to be laid here. There is something very touching about it to me.

I am very anxious to put a wire fence around the spot. I think it must be nearly two acres, and I do not like the animals to have the run of it. Two or three years ago I told the people that if each family would contribute something toward the wire I would put up the fence, giving the cedar posts, and the expense of putting it up. A number brought a quarter each, so that I have $4 toward the wire. Things have been so with me for several years that I cannot make up the sum lacking, so the fence must wait a while longer.

September 15.

My poor, dear Chloe was so excited after old Mose's funeral that she came last night and stood talking until eleven. I was frantic to finish a horrible French play I was reading — it is in five acts, and I am so tired of its wickedness, and I cannot finish it. When I saw she had started on a regular talk I got my sewing and did a lot of work — put the whole frill I have scalloped on to the skirt.

She said the funeral had been grand. Michael had spoken beautifully about "Uncle Mose" and given him such a character; said that white and black respected him and were kind to him, and when he had said all that you would think could be said, Harris had taken it up and praised him more.

"En, Miss Pashuns, dem call yo' name too. Bre' Harris say, en de chile of his ole Marster thought dat mutch ob Bre' Mose dat she sen' fo' tell um say she was gwine to give him, a oxen for him to put in a little cart to drive out wid."

Moses had sold his fine pair of oxen some few years back. With the price of one he paid for having a cataract removed from his eyes, and the rest of the money he asked me to keep for him, and it lasted him four years, I think. He would send his grandson, a little boy, down with a note to ask me to send one, or five dollars, as the case might be. Chloe went back to the time about twenty years ago when he planted rice here, when he had always from $300 to $400 in the bank; but his health failed and he had to move away from the plantation to the pineland because the "tissic" was not so bad in the pineland as on the river. It was very touching to hear all she had to say. She wound up by saying why she was so glad that the minister was absent, for she said: —

"Dis preecher don' keer fu' nuthin' but money — dey ain't no money un a funeral, en he jes' hurry thro' en don' hav' no proper preechin', neider talkin' nor singin', let lone prayin', en, Miss Pashuns, he's a black man too, en dat mek it wuss."

"Why," I asked, "does that make it worse?"

"Bekase eberybody kno' mulatta lub money, en dey look fu' dat, but w'en yu see a daa'k complected man, same color 'bout as me, yu don't look fu' dat, kase dat cola' don' lub money like dat."

That was news to me. I did not know that what she

called "light-skinned" people were more avaricious than black. I was disposed to argue on the subject, but Chloe was emphatic.

"Nigger get dem fault, 'tis true, but dem don't wushop money." With this high praise for her own special color she said good night.

Jim and Joe Keith have had a tremendous day's work bringing in the hay. The ox wagon has a very long, high rack body, and that was packed just as tight as it could be ten times.

My dear little Scottie dog came and sat very close to me to-day and looked at me with very sad eyes and dribbling badly from the mouth. I am too distressed because that is the way my former little terrier dog's last illness began.

Went to Cherokee, taking Chloe, for I was to have the peanut crop harvested and I felt I needed all the eyes possible. Goliah worked finely and it was a successful day. The peanuts turned out so well I had to send for two extra hands to get them all in.

I would feel very proud of the yield if there were not so many "pops" in them. Hypocrites they are. They look perfectly solid and plausible and when you break the shell there is nothing in it. I should have used more lime in the land.

Walked down early into the cotton-field, and found it full of tracks of bare feet, so many different ones that I think every one on the place had been out stealing cotton. Jim was away to-day, so they had no fear of being caught. There is no one else who would tell me if he saw them there. I have felt sure this was going on because I have watched the cotton and know that from day to day it disappears. There is no hope of making anything.

At a store near-by they buy seed cotton in any quantity, paying $3\frac{1}{2}$ cents a pound. Little Jim-Willin', who can pick twenty pounds in a day, can go and sell that amount for 70 cents, and no questions asked. If they pick every morning during the two hours they know Jim is milking and attending to the horses, they can get all the cotton as fast as it comes out. I have wondered because it opened so slowly that I can only pick once a week.

To-night little MacDuff seems very sick. At 9 I get Chloe to go out with me and I put him in the little wire enclosure so that the other dogs cannot worry him; gave him milk, of which he drank a very little, and put a big pan of water for him.

September 16.

This morning when I went to see MacDuff I took him some griddle cakes, which he prefers to anything else. He tried to eat and had a dreadful convulsion. I sent Jim for hot water and bathed him. Then he seemed much better, walked to the fresh water I had brought, and drank heartily. Then he went and tried to eat, but the same thing came on, only not so severe. He has a sore place on the side of his mouth and also on his tail. The flies worry him so I have rigged up a little mosquito net for him.

Dear little Duff much the same. He would not take the milk I brought him nor eat, but he dipped his nose deep into the basin of water very often. I think his jaws are locked, and so he cannot lap with his tongue as dogs generally do, but he can draw in water through the sides, when he buries his whole mouth in the water. It is distressing, to see him suffer. I do not think he is in pain, but he must die if he can neither eat nor drink. I had to try to force his mouth open and pour some milk down, but I saw it hurt him and it did no good. His gums seem very sore and inflamed.

To-night I finished reading "Queed," a most delightful novel, that leaves a good taste in your mouth, which is more than can be said of many.

September 17.

Found my dear little dog dead, when I went out to him this morning, my little silent partner. I will miss him sorely. I thought it best to bury him at once on account of the other dogs, though I was sorry to do it without Jim, who has been so good to him.

Goliah felt most important in having to dig the grave, which was hard work, for I wanted it very deep. It took him a long time. Chloe, Lizette, and I, all assisted. About eighteen inches below this sandy soil is a thick stratum of very stiff clay.

I had some trouble to find a suitable box, but did find one, and then I put him in myself and put his ivory back brush in with him and his mosquito net, then covered him with the soft gray moss and closed the box and lowered it into the grave with ropes. Had Don and Prince in leash looking on. When we had finished filling in the earth, I said a little prayer asking the help of the Good Father to be as faithful to my duty as this little dumb beast had been to his.

I always remember that seventh verse of the thirty-sixth psalm of David, which comes in the Psalter for the seventh day, morning prayer, "Thou, Lord, shalt save both man and beast; how excellent is thy mercy, O Lord." We cannot fathom the mercy of the Infinite One.

Just as I finished Jim came and was quite shocked and distressed. We had thought MacDuff much better when he left yesterday. He said I ought to have waited for him. Chloe, Lizette, and Goliah had made quite a tall mound of clay over him and Jim begged me to let him take it down and rearrange it. Chloe said: —

"I tell Miss Pashuns 'tain't de fashion to hab a high grabe, but him say mus' put all de clay on."

I had scattered white clover seed over the mound, but I let Jim do what he wished and he spent some time over it;

made a neat little mound the shape of a casket, then planted clover over it, and got some plants of petunia I had in a box, and planted them there. At the same time he made up little Zero's grave, which had got quite flat, and planted flowers and clover there.

I made the others rake up all the leaves and trash and burned it. I was afraid the other dogs might catch the disease, whatever it was — spinal meningitis or lockjaw, I think. This ends the chapter of one faithful little unit who did his best always.

PEACEVILLE, Monday, September 18, 8 A.M., 1906.

It has been blowing a gale all night and the mercury has fallen nearly to fifty and I am looking into my trunks to find something thick to put on. Yesterday was a perfect autumn day but with just a little something in the air that suggested a storm and made me name Mosell's splendid new calf Equinox, and to-day there is no chance of having any kind of work done, for though the wind is so high the rain is falling steadily. 11 A.M., the storm is raging and as usual I am greatly excited and exhilarated by it. The voice of the Great Creator seems to be so distinct in the storm; "The floods clap their hands and the waves rage horribly, but the Lord is mightier."

Four large pine trees have fallen in front of the house, and it shakes every now and then as though it must fall. Minty is terrified, but Chloe stands firm, trying to quiet Minty's fright. I sought out a sheltered corner of the piazza and told her to take her work-basket and go on with her darning, as I know occupation is a great soother to the nerves; she seemed much calmer and I went in to follow my own prescription and write, when she came rushing in crying aloud, the immense Water Oak just at the north of the house had fallen, and taken off the corner of the shed. Most fortunately the wind was west of north; the whole house would have been crushed if it had been due north. Even Chloe seemed a little upset at this, so I called them into the sitting room and standing I read the 93d Psalm and then we knelt for a short but earnest prayer and then I said, "Now that we have put ourselves into God's keeping, we need have no fear." The effect was wonderful; Minty did not give way to her terror any more.

2 P.M.

Standing in the piazza it is a strange sight; huge pines falling on all sides; the roar of the wind is so great that one does not hear the falling of the trees. Many are snapped off halfway up as though they were splinters, while others are rooted up, the huge mass of roots standing up ten feet in the air.

The wind is moving round slowly from N.W. to N.E.; if it gets due east blowing at its present rate this house must fall; for there are two defective sills and four blocks, and if the wind ever gets under the huge shed of the piazza to the east and south, it would lift the whole house from the underpinnings. I have told Chloe she must boil some rice on the kerosene heater in the dining room, as it would be dangerous to attempt a fire in the kitchen, and every one is exhausted. If we are to have the house fall over us it is better to be in condition to stand it. Chloe is most reluctant to obey me, she seems to think it will be a disrespect somehow to the storm, but we do not know how long it may last, so I insisted.

Rab has just come in, the wind having lulled a little. I have been very anxious about him as I sent him before the wind became so terrible to Miss Penelope's to get some groceries and ask for a receipt for pepper pickle. He tells me Miss Penelope's house has been destroyed and the ladies had all run out into the storm just escaping with their lives from the falling bricks and rafters. My impulse was to go at once and bring them up here to stay with me; so I started out, and climbed over the great pine which had fallen across the road inside the gate; and when I got on to the public road I found a solid barricade of huge pines across it as far as I could see. I climbed over ten and then felt so exhausted that I knew it would be folly to go on. When I got back Chloe brought me some deliciously cooked rice and a smoked herring, and it seemed to me the nicest thing in the world, though it was a strange thing for her to choose. I suppose she had overheard me talking to some one of the government experiments as to the strength of the different foods, and how they had found salt fish a most nutritious and strength-giving diet; it certainly proved its stimulating qualities on this occasion.

I felt quite refreshed and, as Rab said he could show me a way to go around the trees, I started out again, this time under his guidance. After much climbing I got into the W's yard; the number of prostrate trees was amazing and such beauties. As I passed the Rily house I saw it was a complete ruin; a part of the roof had been blown away about 300 yards.

I found the W's in a little one room cottage which fortunately was on the lot; they were soaking wet, but were so happy to be alive and together that nothing seemed to depress them, though the house is a complete wreck and in such a tottering condition that nothing can be got out. I begged them to come home with me, but they said it was impossible for them to leave, and by the time I had climbed over all the barricading trees and got back into my tottering house, I came to the conclusion that they were wise. They were going to sit up all night in the cottage.

Friday, September 19.

I went to bed and slept soundly last night. I find very few people went to bed at all, as the wind continued blowing heavily all night; but this morning it has lulled and there is a drizzling rain and dull gray sky.

I sent Jim to Gregory Saturday to remain there and bring R. up, as he had written me he would be there Monday. So there was no one at Cherokee but Bonaparte and I knew he must be almost frantic there alone, and as I could hear nothing of what had happened there I thought it was my duty to make an effort to get down to the plantation. So I had Rab put the side-saddle on Ruth this morning and started off, taking Rab along on foot; for though he is very small, he is so brave and intelligent that I knew he would be a great help. He was delighted to go and very proud of the position of escort.

By great meandering around the tops of trees, and jumping others, through people's yards I made my way to the post-office. There I met Mr. B. and Mr. F. who said it was madness to attempt to go down to the plantation on horseback. I said I would tie my horse there and try it on foot; but at that they both agreed it was better for me, if I insisted on going, to ride as far as I could, and then tie my horse and proceed on foot. The road goes two miles due east and then makes a sharp turn and continues two miles due south. About a hundred yards out of the village Mr. R. appeared, looking haggard and weary; he said the storm had been terrific at his place, he had had nothing to eat since yesterday morning, and that the road was absolutely impassable; he had made his way through the woods and it had taken him hours; but he told me there was no hope of getting along the road as the deep ditch on each side and the trees across it made it impossible. I took his advice and struck into the woods, greatly against Rab's judgment, and Ruth fought every step of the way. My hope was to make the other side of the triangle by going in a southeasterly direction, but there was no sign of the whereabouts of the sun and so nothing but instinct to go by. Rab rolled up his little pants as high as he could, the continuous rains having made the woods in many places more than knee-deep in slush and water.

When he reached any specially difficult spot, he would fly ahead and from his vantage ground call to the recalcitrant Ruth, "Cum on, Root, look a' me, I git ober en I sure yu'se bigger den me"; and the mare, who is far too human, and had taken her key-note from my pessimistic advisers at the P.O., would seem greatly encouraged by Rab's optimism.

It seemed to me a perfect miracle when we finally emerged from the woods, just where I had hoped to strike the road, on Cherokee soil. At the gate, standing looking distractedly up the road, I came upon Bonaparte; he looked gaunt and wild, and as soon as he saw me he called out, "Teng God! I look, en I look, en I look, en you ain't come."

"Well," I said, "you may be thankful to see me now, for I have gone through great perils getting here."

The faithful old man walked beside my horse, and told of the terrors of yesterday, how he watched his chance to go from the barn-yard to the house and back again, for trees and rafters and beams were flying in every direction; and how he managed to get the five horses and two colts into the cow stable, and how no sooner had he done it than the horse stable fell, the roof blowing away a hundred yards, and then when he saw the cow stable shake he tried to get them out, but they would not come out, and just then he heard a great noise and behold the big two-story barn packed with hay had fallen and crushed Jackson (an ox) and mashed up all the wagons and other farm vehicles and implements in the shed alongside. And then the shed to the engine room went, and then the screw, for carrying the rice from the thrashing mill to the shipping barn on the river, went. It really was thrilling to hear the old man's graphic narrative, and how he and all the animals escaped, was a wonder.

He led me round through the fields and over ditches, for there were five huge live oaks down on the avenue, some directly across. When I reached the barn-yard I jumped down and turned Ruth loose with the saddle and bridle on her, as I can always catch her and I wanted to walk over the whole scene of destruction with Bonaparte. Every fence was down and the sheep, hogs, cows, and horses were eating up the corn, peas, and potatoes over which we had labored so. We went on to the cotton-field which I had seen on Saturday white with a splendid growth of cotton. I planted a long staple cotton which grew as tall as I am and branched so well that Jim could not give it the last plowing it should have had, because the branches met in the rows. I had it picked twice when some bolls were out, but I was advised to wait until the whole field was out, as the hands picked more and the cotton was better quality; so I waited and the picking was to have begun yesterday. Saturday it looked like a field of snow. Now there was no vestige of white, the stalks all lay prostrate, and the cotton beaten into the earth. It is a terrible blow, as I had counted on this for my money crop.

Before catching Ruth I went to the dairy, skimmed the cream, and put it into a preserve jar and determined to try to take it with me. Hearing me lament over my inability to take the milk with me too, for I have been sending milk regularly to several friends in the village who had none, and I knew they would miss it, Rab said, "I most think I kin carry de milk."

"No, Rab," I said, "not four quart bottles. If we had it in a demijohn I could put it in a sack and you could take it on your shoulder, but not four bottles, they would break."

"Le' me try, Miss Pashuns, I kno' I kin do um."

Of course I was delighted to let Rab try, for if he failed, it would be only the loss of the milk.

There was no horse feed at the pineland, so I told Bonaparte to put half a bushel of corn in a sack with the cream jar and a small quantity in one end and the rest of the corn in the other end to balance, I threw it across the back of my horse and tied it securely to the saddle and we started. I heard Rab say, "Root, yu try for roll but yu couldn't mek um," and I saw that one side of the saddle and bridle was muddy. Fortunately she had tried to roll on the right side, so the pommels were not broken. Our homeward journey seemed to me worse than the morning's. Perhaps I was only tired and discouraged at the wide-spread destruction; but we struck a denser growth and did not find the ridge we had followed in the morning. Then we had gone round the two swamps in a wonderful way, now we had to cross both, and the fear of getting caught in those woods by the night made it worse; however, just as the dusk was falling we struck into the road near the village about 300 yards from the point at which we had left it in the morning, and now I was able to trust Ruth to pick her way round and over the trees as we had come that way in the morning.

As I rode up to the house my heart was gladdened by the sight of R. running down the steps to meet me. It was such a relief that I nearly fainted and he had to lift me down from my horse. He had left Gregory with Jim at 7 o'clock this morning in the buckboard much against every one's advice. But he thought by taking axes to cut the trees out of the road there would be no difficulty. They cut, and cut, and took the horse from the buggy which they lifted over many, many trees; but at last they left the buckboard in the woods. R. led Nan and Jim took the big umbrella (which I had just bought) and swung the two valises on that over his shoulder and they climbed home, having just reached the house when I came.

R. said he was just going to start to look for me, he was so shocked to hear that I had gone to the plantation. It was too delightful to have R. with me safe and sound and to have all the dangers and fatigues of the day past, and to know that for the next two weeks I need not cudgel my poor brain as to what should be done, for he has excellent judgment and will direct the work of restoration much better than I could.

Only one thing I must do myself, I must go to Casa Bianca and see the destruction there and hearten Nat up a bit; of course all chance of rice crop is gone there as well as at Cherokee.

Rab had the happiness of carrying around the milk which he had so successfully brought out and telling of the dangers of our trip and of his powers; and he was greatly praised and complimented on his feat.

Chloe had a delicious supper for us which we certainly enjoyed, for neither of us had tasted anything since morning.

September 20.

Drove down to Casa Bianca to find a scene of great desolation in the Negro street or quarters.

Three houses have been blown down but no one was hurt, not even any animal injured. The barn is also down and the trees around the house badly torn, but the dear old house is not hurt in the least, which is a perfect wonder. I love every old board and shingle in that house and I expected to see it knocked flat, with all my dear old furniture crushed, but positively there was not a pane of glass broken or a shingle off, and the great Olea fragrans which grows as tall as the roof was in full bloom, not a branch broken, making the air painfully sweet. It made me very happy and then the childlike faith of the nigs has a soothing, cheering effect. When I asked Nat about the rice, he pointed to the place where the rice-fields should be, which looked like a great lake, no banks being visible. "All gone, Miss; me en Jonas jis done cut de rice and de sto'm come en carry um right out to sea."

When I condoled with him he answered joyfully, "Miss, I too tenkful I see de sun myself, en me chillun, en me wife, fu' fret 'bout rice. Ebrybody stan' same fashun."

That is the great consolation, it is nothing personal and special, every one shares in the disaster, and to them it makes it easy to bear. For myself it does not help me, it only makes it harder, the loss is so widespread and complete. I fear the storm drops a dramatic, I may say tragic, curtain on my career as a rice planter.

The rice-fields looked like a great lake.

CHEROKEE, September 23.

I did not go to Peaceville to church this morning, C. having written me that the dean would hold service at St. Peter's-in-the-Woods at four, that he would drive him there and that they would come back with me and spend the night. So I prepared a nice dinner of wild duck and broiled chicken for them and took my own frugal little lunch at 2 o'clock and drove out to the church.

The day was very hot but beautiful. Jim did not go to town last evening, but stayed to drive me out in wagon with Ruth and Marietta. I always enjoy driving behind this pair of mother and daughter. Marietta dances along gayly and Ruth tries her very best not to let her daughter outdo her.

The little church was crowded with the simple, pathetic congregation. Louise Moore looked sweet in her black dress, her face so sad, for it is just two months since she lost her beautiful seven-year-old boy. She held up with pride her new baby, which she tightly clasped in her arms. So it is with them always: "Le roi est mort; vive le roi." But I know this new baby can never be as clever and good a child as the little Charley was.

The dean read the gospel for the day, the fifteenth Sunday after Trinity, beginning "No man can serve two masters." He shut the book and spoke quietly and beautifully on what makes a Christian. Is it baptism? No. Going to church? No. And so on, and then he explained what it was; to follow in the footsteps left by the Saviour; to be kind, gentle, thoughtful to others, helpful in word and deed, unselfish, self-denying, and to be guided by counsel from above, asked for daily, hourly, in what we call prayer. Then he explained how one can pray in the midst of work and turmoil; the heart needs no ceremony to ask the help of the Heavenly Father. There are no doorkeepers or guards at the gates. There the humblest kind of prayer, the silent aspiration, "Lord, be merciful to me a sinner," goes straight, unimpeded, to the throne of grace. Then he dwelt on God's special care for each of His children and His knowledge of their weaknesses and temptations and needs.

It was a joy to me to see Solomon's great round eyes fixed eagerly on the speaker, looking for something which his simple mind could grasp and hold. He was asking bread, and I was so thankful that he was not getting a stone. Col. Ben fixed his wistful eyes with their long lashes, the only spark of beauty about his very freckled face, on the preacher and never seemed even to wink. A very stout man in his shirt sleeves whom I did not recognize as one of the regular congregation (the men are all so thin and tall), looked as though he was forced to listen and understand against his will.

C. and the dean were not able to come home with me to spend the night, to my great regret. They found at the last moment that it would be necessary to return to Gregory this evening.

Monday.

I sat and sewed, not feeling up to much exertion. I finished the gray muslin frock and it is sweet. Then I darned stockings. If anything can make you more conceited than darning stockings I don't know what it is unless it be early rising. I think if by great and sustained effort I ever became an early riser I would become insufferable. All my life that has been my greatest ambition, but I never succeed in rising early for more than a few weeks at a time under pressure.

September 25.

Danton was to be married this evening and specially invited Chloe, and I saw she wanted to go, so I told her she must go, that it was only due Danton, as a refused suitor for her hand, that she should attend the wedding. She said she did think it would be a "good polish" on her part to be present, but that she could not bear to leave me alone from early in the afternoon until late at night. She said:—

"Dem say de weddin' gwine be at fo', but nigga fo' mean buckra seven! Dem neber will reddy by fo'."

However, I insisted and persisted until she went. I kept Patty Ann as long as I could by going to the barn-yard to measure the peas and telling her she must take charge of the house and yard until I got back. There were about twenty hands. Some wanted to pick for one-third of the peas, some for money. It is very hard to divide peas evenly. I find the easiest way is to weigh the peas and then divide the weight. It really takes less time than any other way, for it is accurate and I can get through weighing and dividing up 1500 pounds of peas among twenty hands in no time.

When I got back from the barn I had to let Patty Ann go. Then I went to see after Pocahontas, my beautiful young cow, who is ill. I gave her a dose of aconite to-day and had Jim rub her with liniment. She is strangely affected; her fore legs seem almost paralyzed and cross each other when she walks, or rather steps, for she is afraid to walk more than a dozen steps at a time; then down she goes.

Her countenance is bright, however, which makes me hopeful. It would be a dreadful blow to lose her. R. L. A., who knows about cows, told me not to take less than $65 for her in the spring before she had her first calf. The calf is so fine I have named her Beauty.

I gave Pocahontas a large bundle of fresh cut grass and a bucket of water, for Jim has gone to the wedding too. To my delight when I put the grass down a little beyond her reach, giving her only one handful to taste, she got up with great difficulty, resting some time on her knees, swaying so that I thought she must keel over, but at last she got to her feet and stood trembling for a little before she could make the one step necessary to reach the grass. It seems to me she has had a severe blow on her left shoulder and the left leg is almost paralyzed. She puts all her weight on the right.

When I had got through my ministrations to her and came out of the stable and started toward the house I was scared almost out of my wits by a tall man standing at the gate. It took all my courage to force myself forward instead of retreating rapidly to the shelter of Pocahontas's stable. However, I went forward with a bold air to find a variation of Don Quixote's windmill. I had been picking up walnuts as my first amusement and got so warm that I took off my white flannel coat and hung it on a tall post near the gate as I went out to see after the cow, and had entirely forgotten it.

I laughed at myself, but feebly, for I have not had such a fright for a long time. There is no human being within three-quarters of a mile of me, and I have had trouble with a very bad man whose hog was in my corn and peas, for a week before we found it out. I had Jim catch it and shut it up and send word to the man that he must pay me $2 before he could get it. That was not nearly the value of what he had destroyed, but I thought it was all I could ask. He sent word he would not pay it, that I could keep the hog and send him a dollar. It is a complete razor back and I don't want it, but Jim seemed to think he would like it and told me he would pay the dollar and take the hog.

I said, "Then where is my pay for the damage to the crop?" It seemed very hard for him to see the justice of this.

In the meantime I am feeding the creature on my precious corn and the owner is very angry.

I put up the walnuts I had picked, four dozen. The squirrels steal them as well as people, so they have to be put in a close place to keep them.

When I went out first the sunset was wonderful. Exquisite rose-colored clouds, layers upon layers of them, filled the heavens and the pure, cold crescent moon emerged from their billowy depths with its bow of hope. I stood and watched with much delight until, oh, sadness! the rose-colored clouds turned to ashes of roses and I felt with a pang that night was near.

When I came into the house at last it was quite dark and I had forgotten to put the matches on the table near the door, so I had rather an eventful progress, falling over chairs and stools to find them. I do not like the dark and cannot walk straight in it. At last, however, I got the matches and lighted the lamp and took my tea, which had been made early but kept hot under a cosey.

After getting through to my surprise I found myself nervous. I do hope my nerve is not going to fail me. It is the first time I have been quite alone in the house and yard without MacDuff, my stanch little watch-dog, and it makes a great difference. He used to sit right by me and follow my every motion, and when he barked it meant something. Don barks because he is chained and Prince barks because he is afraid of the young moonlight, it is so mysterious. Their continual barking worries me, for Chloe left both chicken coops unlocked and I cannot discriminate between the bark that means something and the other.

I tried to read but did not succeed, so I took to the piano and worked hard at Chopin's études. The "Revolutionary" étude requires plenty of work and effort and concentration. Now it is 10 o'clock and I had to stop for very weariness and am writing.

After going out into the yard several times to see if Prince was barking at any reality, I brought him into the pantry, and in order to compose his nerves I left the "Revolutionary" étude, which is so tempestuous, and played the next, which is almost a lullaby, so soft and soothing is it, and to my relief it had its effect, and Prince went to sleep.

Just now I heard Don give a real bark and went out to find the party had returned, Chloe in high glee. I had sent by her to Danton as a wedding present, a white waistcoat. The buttons had been taken out when it was washed and had not been put back, and when I thought of the waistcoat as a present, Chloe was all ready and I could not put the buttons in, so I wrapped them up and put them in the pocket, wrapped the waistcoat in a paper, with my good wishes on the paper, and gave it to Chloe.

She told me that they met the bridegroom on his way to the wedding with his four attendants and she delivered the package. Danton opened it, expressed great delight, took off his coat and put on the waistcoat, and the groomsmen all assisted in putting in the buttons with the string which tied the parcel, and the bridegroom was resplendent and the four groomsmen were loud in applause. He folded the paper up and put it carefully in his pocket and said, "I gwine to keep Miss Pashuns's good will she write on dis paper."

Chloe says the wine flowed freely and the large company all had some, and the six cakes were not all consumed. She is telling me every detail of the gayety. Jim had the duty of handing round the wine, which Chloe privately told me was very sour. But it was a "high class" affair, not so much as a "cuss word," all pleasant and polite.

I am writing while she narrates, which will account for incoherence. She stands with her hand on the knob of the door and every now and then my hopes rise as she opens the door, but at once some new detail comes to her mind and the door closes, the lantern is put down, and the wonderful witticisms recommence.

The bride wore yellow satin with two skirts, a long veil put on with a wreath of white flowers and looked "truly han'som, en I neber bin to a freer weddin', Miss Pashuns, dan dis. Of coa's de cake was to dem tas'e, not to yo'ne, but dem had a plenty o' dem, sum lef'."

At last Chloe has said good night. I'm going to bed to dream of the wedding.

September 26.

Nice letters by mail to-day, but I am so down and miserable in spirits, that I do not know what will be the result. I feel ill just from despair. There seems no outlook anywhere — no hope. I feel ashamed of myself, for there are so many so much worse off. God forgive me.

CHAPTER XIV

September 28.

TO-DAY'S mail brought me a most agitating letter from dear C. She wrote from the hospital, where she was to be operated upon for appendicitis that day. She begged me if possible to go at once and take charge of her household and three little boys until her return. It seems almost impossible to leave just now — but I have determined to drop everything and go. I feel very anxious, for her letter took a long time to reach me.

UNITA, *September 30.*

Left Peaceville yesterday at 3 P.M. and reached this flourishing town at 12 to-day. The boys are splendid fellows, aged eight, seven, and four years, full of life and fun and chatter; it is a great contrast to the silent home I left.

On the journey up I had two hours in Columbia, which I spent with B. I was very pleased to find that W., her ten-year-old son, had recovered from the effects of the great shock he had when I was there in June.

All the children were at a large picnic on the outskirts of the city in a very pretty spot with a stream running through which opened out into a small lake in one spot. There W. was playing with a comrade of his own age, to whom he was devoted. They were wading in what seemed a sheet of shallow water and were throwing up a lemon as a ball, each trying to catch it.

Suddenly as the friend leaped to catch the ball he sank from sight in unsuspected deep water. W. saw him rise and sprang after in an involuntary impulse to save him. He too disappeared. Both rose, then sank again, W.'s second sinking being the friend's third, his arms being clasped around W.'s waist.

About 300 yards away a young matron who had brought her four children to the picnic, one being quite a baby still, heard the cry and started toward the pool at a run. She reached the water as W. rose for the third time. Though panting from the run she sprang in where she saw him sink, and after what seemed to the onlookers an age she appeared holding W. by the collar, and slowly and painfully dragged him to the shallow water. A lady seeing she was nearly spent, waded in waist deep and helped her bring him to the shore, and said: —

"Come out, you are exhausted."

"No, no!" Mrs. M. answered. "There is another, I must go back for him," and she turned again to the deep water.

But every one saw that she could not possibly go down again, and she was pulled gently to the shore and placed in an automobile and taken home.

It was a most heroic action for that frail young woman, exhausted from the run, to plunge in with clothes and shoes on. I asked B. to take me to call on her and that visit will always remain in my memory as a beautiful picture. She was sitting on the vine-covered porch at her sewing-machine, while the children played around. She did not wish to speak of the tragedy and talked of lighter things.

She told us she had grown up on a plantation near Beaufort, and she had only consented to come to the city on condition that her home should be on the outskirts, where she could have large grounds with flower and vegetable garden and keep a cow, a small farm in fact, and her present home gave her all the country occupations and pleasures, while her children could reach the city schools easily. I had gone for the purpose of expressing my gratitude and admiration for her presence of mind and heroism, so before we left I spoke of the tragedy. Her lovely face went white at once, and she said: —

"Oh, but the other boy! If only I could have got him too! I think of that other mother always."

My very heart was stirred by the heroism of the whole thing, and the mercy of the rescue of my dear great-nephew and the terrible tragedy of his companions' death. Altogether my admiration and reverence was excited for all the actors in the drama, Mrs. M., and my great-nephew, who had tried to save his comrade, but Mrs. M. above all; — just a flash of tragedy, heroism, and nobility out of the clear sky, when often life looks so commonplace.

October 4.

It is a wonder to me how I can cast away all thought of things at home so completely, specially the pea-vine hay, which had become a kind of fetish with me, but truth to say, my time and thoughts are so fully occupied that I have no chance to dwell on anything outside of these four walls. It is time for the incubator to hatch and I find myself in spite of active occupations wondering as to how Chloe and Patty are getting on. Neither of them seemed able to make out the thermometer, so that my kind friend Mrs. S. had promised to go in once a day and look at it. It will be a wonder if any chickens hatch under the circumstances.

This is a lovely place, a charmingly comfortable house and so surrounded by trees that I can forget I am in a city and only feel the comfort of having the butcher and baker and every one else you want, call at your door. I certainly can appreciate that part of city life.

"Seven" and "Eight" go to school and the excitement of getting them off in the morning is intense. They have a long walk, entirely across the town, and there are many snares and pitfalls in the shape of circus pictures on the way, so that it is necessary to give them ample time to get there before the bell rings.

"Four" stays at home and is my constant companion. If he were not a fascinating child it would be very trying, but besides being strong, healthy, and handsome he is perfectly obedient and very original. I was sitting on the porch sewing and he was playing with my trunk strap, greatly to the injury of the strap, twisting it around a tree. I told him to stop, which he did at once, and with the greatest agility, to use his own word, he "skinned" up the tree. He went up until he was on a level with the second story windows and then began to discourse.

"Aunt Patience, did you see how quick I minded you, an' stopped doing what you tol' me not to do?"

"Yes," I answered, "and I was greatly pleased."

"Well," he went on, "I did that because if you do what grown-up people tell you not to do, God don't like it, an' he'll surely make you stump your toe."

I could not help laughing, it was so funny, his little bare feet are so battered and bruised by stones and roots; but he got very angry, and let go his hold on the limb on which he sat, to gesticulate fiercely as he went on.

"That ain't nothin' to laugh at, its puffectly true! An' if you're a grown man an' won't mind Him an' do wrong, He might make you break your neck, but if you're a boy, he'll only jes' make you stump your toe."

I was so afraid that in the earnestness of his gesticulations he would fall from the high limb that I became solemn at once and said how pleased I was that he was so wise and realized that evil doing always brought its own punishment. At the same time I begged him to hold on, as it was necessary to take proper precautions not to get hurt as well as to do what was right. As "Seven" fell from that same limb and split his tongue two days before I arrived, I was truly thankful when the little preacher got down safe and sound.

The last time the boys stayed with me at Cherokee Chloe nicknamed them the doctor, the lawyer, and the preacher, and the names seem to suit. I was walking with "Eight," the doctor, yesterday afternoon and as we flew along, for I walk fast, he threw his arms out and exclaimed: —

"Oh, I just wish I had all the money in the world."

I was quite shocked. "Oh, my dear boy, what makes you wish for money? You have everything you want."

He answered: "Didn't you see that poor old daddy, all ragged and dirty? He has an awful foot, I saw it, and I gave him a dime the other day, but if I had all the money, I'd load up my pockets with big bills and as I went along the streets and I saw him, I'd just slip a fifty-dollar bill out of my pocket and into his hand and say, 'Shut your hand quick, old uncle, here's fifty dollars; go get your leg cured and buy all you want.' And then I'd run on quick before he knew who it was. And you see that poor, thin, pale-faced little girl coming out of the factory? I'd do the same to her, and walking just as fast as we are now I'd just give everybody that looked needing it, a good big bill! Now wouldn't that be jolly? And wouldn't I be happy!"

I told him if he ever wanted to do that he would have to work hard at his arithmetic, over which he has so much trouble, for there was no chance of ever making headway in the world without conquering that — which seemed to put arithmetic in a new light to him.

But I really was very pleased to see in the boy that love of humanity which made him wish to relieve suffering, though only in imagination, instead of dreaming of autos and other grandeurs for himself. "Il chasse de race." But we certainly understand spending money better than we do making it, which is a pity and made me point out to him that money had to be made before it could be given away, and that money was made by arithmetic, so to speak, rather than by dreaming.

October 20.

I have got on beautifully with the boys and am so happy to know them well. I have had many trials of strength with them, but I never give in. The "doctor" came in from school the other day and threw his arms around me and said: —

"You are just the sweetest aunt in the world!"

I said, "What does this mean?" laughingly, for we had had a mighty tussle that morning over his arithmetic.

He went on as if not hearing me: "I just get praised in school all the time, since you have been here."

I thought it was the most magnanimous thing, for I had been very severe on him in the battle over the arithmetic. I really think the mental arithmetic is quite too hard for a boy of eight, it requires such an effort and so much concentration; but as the lesson is given him, he must put his mind on it and learn it.

The analysis is more puzzling than the questions themselves, and he fights it, and I don't wonder, but as the lesson is given it is his duty to learn it, and I make him shut his eyes and concentrate his mind; and I thought it was wonderful that having felt the good result in praise he should wish to pass it on to me. Oh, the joy of having first class material to work upon!

CHEROKEE, November 10.

A perfectly exquisite day. I reached Gregory last night and spent the night by invitation at Woodstock. I had written for Gibbie to meet me there at noon, and he arrived punctually. I rested the horses about half an hour and then started back. The horses looked jaded and I let them walk, as it was intense enjoyment to feel the soft balmy air on my cheeks and to study the variety of lovely wild flowers which autumn brings, as we went through the pine woods, following the rough and winding short cut to the ferry.

I asked Gibbie questions, to which he gave the most prolonged and elaborate answers. I am sure he had composed and arranged them all as he drove down. He told me every item of home news; everything rose color; potatoes dug and very fine, "about t'ree hundred bushil." "Great crop of hay," he having saved it all. More peas than I "could 'stroy." He spoke of Bonaparte altogether as "the Cap'en," which showed me they were on good terms, a most unusual thing. All the cows in fine condition, he reported, but when I asked about Heart, the Guernsey heifer I was so anxious to raise, he said a sad accident happened and she was dead.

After we crossed the ferry the horses looked so downhearted that I asked if Ruth had had any holiday. No, he said, Ruth had been driven every day; but Romola had done nothing since I left home.

"What," I said, "not been in harness since September 20?"

"No, ma'am, just been out in de fiel' de eat grass."

"Mercy on us," I cried, "and you brought her on this twenty-six mile drive to-day?"

"Needn't to fret, ma'am, Uncle Bonapa'te feed um well, he give um twenty-eight year o' co'n jes' fo' we sta'at dis mo'ning."

I began at once to feel anxious about Romola and drove slower and slower. She would turn and bite at her side from time to time and travelled with her head down. Finally when we were five miles from home she threw herself violently down. Romola is such a good creature; she managed not to break a strap of the harness, nor the pole, only she nearly toppled Ruth over, but by falling against her she saved the pole.

I sprang out and had the harness taken off quickly and got her up and led her out of the road into a grassy place in the woods just in time, for she threw herself down again, rolling over and over and groaning and tossing herself about — a genuine case of colic.

First I told Gibbie to run home and bring Nana as quickly as he could. Then I considered that it must take at least two hours, even if he ran, which I knew he would not do, and to be left on the highway alone, the buckboard loaded with my possessions, with a sick horse, would be a trying ordeal for me and would really be tempting Providence in the way of tramps, so I said : —

"No, don't go yet a while; perhaps some one will come whom I can send."

In about half an hour a neighbor passed and offered to help me, so I asked him if when he passed Cherokee he would drive in, and tell Bonaparte to bring Nana in the old buckboard and to tell Chloe to send by him the horse physic from behind the dogs on the mantelpiece in the dining-room. He seemed very glad to do it and I felt relieved, knowing I would not have to spend the night on the road. I always keep a bottle of aconite behind a very beautiful pair of bronze hounds by Isidore Bonheur and Chloe knows just where to find it, for I have kept it there for years.

Romola continued in great distress. I had a bottle of almond oil with extract of violet in my valise which I fortunately thought of. I got it out and told Gibbie to rub her, but finding that he didn't seem to know how to rub, I just took it myself and rubbed her well. I had to be quick in getting out of the way when she flopped over or I would have got mashed; but I stood behind her and, leaning over, put my whole weight on my hands.

As the sun was dropping below the horizon in the west she got up and shook herself. I led her about a little and felt sure the attack was over, so I told Gibbie to harness and put her in. He had been kept busy by Ruth, who as feeding time approached was eager to break away and get to her stable. We moved off slowly, and in half a mile we met Bonaparte with Nana and the other buckboard and he took the trunk. I gave Romola a dose of aconite and she plucked up a little spirit; but she did not pull, she simply walked beside Ruth, who took the whole load.

It was after dark when we reached the house. I gave her three quarts of hot water with soda in it and another spoonful of aconite. I was truly thankful when I finally dragged my weary limbs up the front steps and found a bright fire, nice supper, Chloe, Don, and home.

Sunday, November 11.

My blessed mother's birthday. I am too stiff and ill to attempt to go down with flowers to her resting-place as I usually do; a great disappointment. Bonaparte asked for a private interview, so I went to one end of the piazza, though there was no one within hearing. He told me after a long and mysterious preamble that he was engaged to be married.

I was distressed when I heard he had selected a comparatively young woman from Gregory. When I expressed my anxiety, saying a woman from the country would suit him better, he said that when I saw Jane I would have no objection to make, as she bore a fine character with white as well as black. Of course I can do nothing now but give him my good wishes.

Two or three months ago when I saw his restless, miserable frame of mind, I knew he was thinking of replacing his good, faithful wife and I tried to help him. After a careful survey of the matrimonial field, I concluded that good little Jinny would be the best person for him. She is an industrious, smart woman, who had been a faithful wife and mother and is now a widow. One day I said to him that whenever the time came that he felt he needed a companion I thought Jinny Robinson would make him an excellent helpmate. To my surprise he answered quickly, "Jinny too old for me, Miss Pashuns."

She is twenty years younger than he is. My mother was always appealed to for advice and suggestion by those left desolate, and I never knew an instance when her selection was rejected or the match turned out badly, so I was quite unprepared for this rebuff.

Jinny lives on her own farm and all her children are married, so that she would have suited him well.

Nat came up from Casa Bianca to tell me my fine yearling steer Knox was dead. He was perfectly well apparently when Jim went down there last week. It always is a trial to talk to a negro in such cases. I asked of course what ailed the steer. Nat scratched his head violently and answered:—

"Miss, 'e time cum, I t'ink. W'en we time cum we 'bleeged to go. De black steer time cum en I cudn't keep um; en beside dat 'e had de hollow tail."

Of course I retreated from the effort to find out anything, but I told him he must bring the rest of the cattle up here. The pasture being very fine down there, I leave the cows there in summer, but as soon as the corn-fields are open here I bring them back where I can look after them during the winter.

CHEROKEE, November 12.

Great activity prevails in this household; I am moved to brush up my dear old home a little, so I have bought some kalsomine, and every minute which can be spared from getting in the hay Jim is kalsomining. He has finished the breakfast room, which was a disgrace, and then he finished the upstairs hall, and is now engaged on the lower hall.

The dear departed peacock, whose mate was eaten up by a fox while sitting on a nest of beautiful eggs, lived three years in a state of single misery, during which time he broke every pane of glass in the windows he could reach. It was so pathetic that I could not give way to wrath and have him beaten away. He was looking for that lovely mate, with her graceful long neck and dainty small head, and seemed to think she was imprisoned in the house, for he roamed round and round it, first on one shed and then on another, and when, peering in through the window glass, he caught sight of his own iridescent form, he would plunge forward in an ecstasy of joy, break the glass, cut his poor, proud head and hastily fly away, only to begin the search again as soon as his wounds were healed. In this way the windows were broken one by one, and the dirt daubers (a very busy flying thing that looks exactly like a wasp, but does not sting) came in and made their wonderful clay houses in the halls, so that it looked like some old, deserted, haunted place.

It was impossible to get all the glass put back. The hall window has a pointed arched top like a church window, and that shape of glass I could not get, so I just felt helpless and hopeless while the little workers in clay triumphed over me. When, however, the hall window was covered with fine bronze wire on the outside from top to bottom these little wonders of industry and perseverance were foiled.

It was funny to watch them when they first reached the window loaded down with red clay and flew up against the wire. They could not believe that that pygmy man whom they had got the better of for years had really foiled them at last. For days and weeks they continued the attack and many, many perished in their determined efforts to squeeze through openings too small for them.

But to return to the peacock and his search for his lost love. He reminded me of that tragic scene in Gluck's "Orpheus and Eurydice" when he dares to enter the vast terrible kingdom of the dead in his search for his beloved —

the eager, pathetic gaze into the lifeless face of each veiled form, the joy of imagined recognition, only to fade into disappointment and horror as returning life shows the mistaken identity. The peacock grew thinner and thinner. Occasionally he would go into the busy poultry yard and spread out his beautiful fan and salaam to the white Leghorn hens and win their cackling admiration, but those exhibitions became rarer each year, and finally he disappeared. I am quite sure he sought the solitude of the forest to die.

I miss him all the time in spite of his mischievous activity, for he was a part of the place. I tried very hard to get a mate for him, but never found one. In the years gone by peafowl were very common through this country. We used to call it our episcopal dish, for every year when the Bishop of the diocese stayed with us on his visit to the parish, mamma had a roast peacock as part of the dinner. The breast is very large, like that of a partridge, and of a very delicate game flavor.

Since the clay workers can be kept out it seems worth while to destroy all traces of them and have the wall white and fair once more. As the work progresses and the air of desolation is subdued my spirits rise, and I wonder how I have stood it so long. It is well there is something to cheer my spirits, for the financial outlook is appalling.

The storm-tossed crop is hopeless, the corn all damaged; in the little that there is, not a perfect ear.

November 14.

Great rejoicing! To-day's mail brought a letter from J. L. H. saying she would be here Monday. Day spent in trying to get the house in winter trim, for it is very cold. Got down most of the carpets and rugs, but could not get the curtains up. They are all sewed up in homespun bags in the spring with camphor or moth balls, and it really is a day's work to get them all out and beaten and aired.

November 15.

A tremendous day. Took my dearest J. down to Gregory and then to Woodstock, where it was so pleasant that I lingered too long. I had had a great deal of business to see about in Gregory, so that we were late getting in to Woodstock. I had fortunately bought a lantern and I needed it very soon after leaving Woodstock, the road being very winding and intricate for the first three miles.

At the ferry the man called to me to drive in quickly, as there was a tug coming down the river bringing him a new flat, and he must get me over as quickly as possible to return and change flats. Goliah, who had gone behind on the buckboard with me, was much excited and added all his strength to the two men in pulling the flat over. When we got about halfway over Moses, the ferryman, saw that the tug was coming down rapidly upon the wire. He called to the captain, a negro, to stop. This did no good at all. On, on, came the snorting tug like a relentless fate.

It was a dreadful situation, for I feared Ruth would turn her head, and then I knew she would jump out of the flat. Fortunately I had driven her hard and she was thankful to be quiet. While I was wondering that Ruth was so quiet something happened, I did not know what.

The four men and Goliah, who were pulling, were thrown to the floor of the flat, Ruth was nearly thrown flat, and the steel rope with which the flat was pulled lashed round the buckboard's wheels, fortunately not reaching the mare. I was thrown out on the wheel and before I had righted myself Goliah picked himself up and flew to Ruth's head, which I thought a wonderful evidence of fidelity to a responsibility, and it was lucky that he did so, for as soon as she realized that we were out of our course and not making for shore, Ruth became very restless and impatient.

Moses yelled to the man nearest the broken end of the rope

to seize it, which he did, and that was a mercy, for if it had escaped we would have drifted down the river without any means of regulating or guiding our course. Then the men all pulled together on the rope and ran the flat up into the bushes about 200 feet from the slip where the flat lands.

It was quite dark, only two lanterns being in the flat, mine and the one the ferryman had. After tremendous effort they got the flat to touch the slip at one end, leaving about four feet of water at the other. I saw that was the best they could do and that it could only stay so for one second, so I called, "Hold it so for a minute." I told Goliah to let go Ruth's head and spoke to her, and as she hesitated gave her a sharp cut with the whip. She leaped out over the gap and we were safe on land.

I drove home too thankful for words for the great escape. Just the thought that we might have been swirling round, drifting down toward the sea with the current, made the drive home seem a delight. When I went to get out of the buckboard, however, I found I was a rag and could scarcely stand.

November 17.

Went out immediately after breakfast and saw Gibbie put half a bushel of cow-peas in the big pot, fill it with water, and make the fire under it to boil food for cows. Then I told him to start ploughing in the half acre of oats. Bonaparte had already scattered the seed. Later in the day I walked down to the field to see the work, and found Gibbie had not done a stroke, had simply gone home, leaving the oats on the earth for the birds to devour. I was too angry to go in pursuit of him. I find it very unwise to speak until I have cooled off. As Bonaparte said to me once, "Ef you don't tek keer dese peeple'll mek yu los yo' soul."

The corn has been so stolen that there is scarcely a fifth of a crop — all the big ears gone, leaving only nubbins. The horses are all weak from lack of food and I feel desperate. Even Goliah is changed! All the joy and fun and play seem to have left him, and his fat little black face looks like a thunder-cloud.

His household at home are urging him to demand more wages, and he does not wish to do it, and yet the clamor there makes him discontented. I brought him a suit of clothes and a pair of shoes, in which complete outfit he sleeps. The weather being very mild, I beg him to save the shoes for cold weather, as he has never worn shoes before; but in vain, unless I put them up for him, which I do not wish to do, for I wish him to have the full enjoyment of them. He found an old pair of white kid gloves in the buckboard. I had used them to wear about the place and somehow left them there. Yesterday he asked me for them and now he wears them all the time — cutting wood, eating dinner.

I tried to translate in concise and striking words the French proverb, "Chat ganté n'attrape point de souris," but it had no effect; he sits gazing at his shoes, his white-gloved hands folded in his lap. I have sent him to school, as the public school is only half a mile away, and there is a good teacher, but nothing can restore the little gay Goliah, who jigged as he walked.

He has eaten of the apple and been driven out from the Eden of childhood and from henceforth will always be wondering how much he can get out of me. I knew it had to come, but I am so sorry, and I miss the little boy so much.

Sunday, November 20.

Our rector's Sunday with us. He gave a very interesting sketch of the church convention, which he had attended. I had to play the organ as well as do all the singing, as Miss Penelope was not able to come. I thought it was impossible, but really nothing is impossible, for when I got home feeling like a rag I found Zadok waiting for me — "to ask my advice," he said.

I had not laid eyes on him for years, but he is the son of a faithful servant and was born in our servants' hall just before the end of the war. He has prospered, married well, and has a large family, who all help in the cultivation of his farm of twenty acres. Now that he has got into trouble he comes at once to me.

His great snare is the dreadful firewater. He told me he had got into "a tangle." Coming back from Gregory one day when he was "not quite himself" he had been accused of cursing and making a disturbance. He had been notified that he would be indicted, and when he went to remonstrate the man said if he paid $25 he would drop the case; but he had put off and put off, and now hears the case is to be tried to-morrow.

I made him tell me everything and felt sure the only thing for him to do is to pay the money as quickly as he can, for if the trial comes off it will go hard with him. He is known as very obstreperous and noisy when under the influence of liquor, though peaceable and civil otherwise; so I told him to get the money as quickly as he could and try to pay it before the trial came up. I was greatly worried about it and had Jim put Alcyone in the small buckboard and drive me down to Mr. B.'s. I took a very pretty apple geranium as an offering to his wife.

I told them I had come to see what I could do to help Zadok, that I was much distressed to hear he had misbehaved, that his father had been our trusted and faithful servant during the war when there was no man at home, and I begged them if he promised to pay the $25 to drop the case. Mr. B. said he thought it had now gone beyond his power to drop it, but he would try what could be done. On the way home I met Zadok in his buggy, which is a very nice one, driving his horse, which is a very good one. I told him I had done all I could and begged him to make a humble apology. He said he would and seemed much impressed.

Little Alcyone is a swift little filly and went splendidly, but she is not large enough to take two people in a buckboard that drive of sixteen miles at the pace she likes to go. It is an unwise thing to let her do it. She should have a light road cart only, for she will not walk at all. I really felt when I got home as though I had been actively employed from the time of Noah and the flood. Zadok promised me to stop drinking. God help him to keep his promise!

Casa Bianca.

CHEROKEE, November 27.

Before I started out this morning I called for Bonaparte and showed him a large portfolio of engravings and prints and told him to make a light wooden frame into which I could slip it to send by express. I specially told him to leave one end open so that I could put the portfolio in myself. When I got back from Casa Bianca this evening I found the very neat little light frame and was delighted, until I found the portfolio was nailed up in it so securely that I would have had to break the frame to get it out. It was too provoking, for I had not meant by any means to send all its contents.

The time for the payment of taxes has come. Mine are over $100 and my little cotton crop cannot cover them after paying my yearly accounts, so I must sell something, and I decided to send some of the things in the portfolio on to New

York to be sold. There are a lot of queer old things in it which many would call rubbish, but which I delight in — a map of the city of Charleston in its veriest infancy and engravings of a horse and dog which had won beautiful silver prizes in 1760 or thereabouts. These things are of great interest in the family and especially to me, who live in the past so much.

Then there were some water colors I wanted to keep. Altogether this seems the last straw to a very tired camel. Bonaparte had gone home, it was late Saturday evening, and Jim is to take the box down to Gregory when he goes tomorrow to send off by express Monday, and I just gave up and let them all go. The eternal struggle against contrarieties and difficulties is too much for me.

The time for tax paying has nearly passed and if I do not send the things off to-morrow they will have to wait another week.

December 7.

The pace has been most rapid for some time and I find that when I have pleasant companionship I neglect my faithful dumb confidant.

Bonaparte's wedding preparations have caused me much anxiety. I promised him some money for the occasion and sold two of my precious young heifers to be sure to have it, but there has been some hitch and the money has not reached me, and when he came for it yesterday I had to tell him I did not have it, which hurt me.

To-day by mail arrived two large packages, both with special delivery stamps on them, a beautiful frock-coat and waistcoat. My dearest L. had sent them to Bonaparte. I hurried out to his house with them. He was out, but I put them in his daughter's hands. He had confided to me his anxiety as to his wedding garments, as he said the wedding being in town he wished to be suitably dressed. I wrote all this to L. and she truly did the impossible in getting them here in time, as she only returned to Washington from the mountains two days ago.

Just before dinner D. arrived, having brought a present of two mallard ducks. I was so charmed to have them, for C. leaves, to my regret, to-morrow. Later I received the check I had been expecting, so that I had the satisfaction and relief of sending for Bonaparte, who had not yet started for Gregory, and fulfilling my promise.

He was radiant. All's well that ends well, I suppose, but I have really suffered from the tension of fearing that the faithful old man was going to be disappointed.

December 9.

Sent C. and her delightful little Albert down to take the train yesterday. Their visit has been an unalloyed pleasure. Jim, being in Gregory, attended Bonaparte's wedding at 8 o'clock last evening. He says it was a most elegant and well-conducted affair, with an abundance of good cake and wine, and his respect for Daddy B. has risen immensely, which is a comfort and relief to me, for they are not very friendly. Bonaparte boasts, "Yes, Jim kin mak de crop, but Jim ain't got de key; I got de key," and of course that is aggravating.

I am very sensible of the thorns which accompany the roses of faithful service with both Chloe and Bonaparte, but all the same I thank the good Father for the thorns, because of the roses which sweeten my life.

December 13.

A good steady rain last night, thank God. There was a frightful danger of fire getting away in the woods. Every one was nervous about it. I am very tired, for I had to burn the chimneys, which always scares me terribly. I always have Bonaparte to help, and this morning he did not want to do it and said there was not rain enough and was rude to Chloe when she called him, so I determined to burn them myself.

I take a large newspaper, pour about a tablespoonful of kerosene on it, holding the end in the tongs, and as soon as it takes fire thrust it as far up the chimney as my arm will allow, which is not far; that is the reason I like to have Bonaparte do it. In an instant the whole chimney is ablaze, with a terrific roaring. Patty was stationed outside to see if any of the blazing soot lighted on the roof; I on my knees in front of the fireplace prayed with all my might; the terror that there may be a crack or flaw in the chimney is always with me. The great matter is to burn often and then there is no great accumulation of soot.

It has always been a thing I had to wind myself up for, and requires all my will to make me do it. Once I was so demoralized about it that I got a man to come with a regular little chimney-sweep from Gregory. The little fellow went up one chimney, but when he came down he wept and pleaded so not to go up again, saying the chimneys were so long, three stories, and the flues crossed so that he could not breathe, and I would not let the man send him up again. It was an expensive experiment, and I concluded the old time way of burning out was the best, and try to make myself do it once a month.

CHEROKEE, December 14.

A July day, rainy and hot. War in the kitchen zone. Goliah roused Chloe's ire and she fell upon him with fury. When I went down to remonstrate with Goliah he was in a great rage and I heard him mumbling, "Yes, if I only had my axe I wud 'a' settle 'em," and nothing I said could have any effect, so I had to tell him to leave the yard and not return.

It will be a loss to me, because I know all his faults and can generally meet them, and he is very competent for one of his size with the horses, and drives very well. I take him everywhere with me on the buckboard, but he has a morose, morbid temper and he has been very rude and impertinent to Chloe, so I cannot keep him.

December 15.

While I was wondering how I was to manage, Jim being away, without Goliah to put the horse in for me to go to church, and again in the evening for the rector, who is here on his monthly visit, to go to St. Cyprian's, the negro church, Chloe came in and said : —

"Goliah is yere. He dun ax my pa'don en I gib um, en I tell him I keep nuthin' agenst him."

I think it was as great a relief to her as it was to me to see him installed again. I trust now peace will reign for a while.

December 16.

Have the great pleasure of our Bishop's yearly visit. He came last evening after holding service at St. Cyprian's, the Colored chapel. It rained all night and looked very dismal this morning, but as we sat at breakfast the sun came out and we were all rejoiced. It is the event of the year in the parish and the disappointment of a rainy day would be intense.

We started for Peaceville at 10 : 15 and I was delighted to find Miss Penelope able to be at her post at the organ. It was a solemn service, with confirmation. The Bishop, Mr. G., and I were invited to a delicious lunch before going on to the little chapel in the woods. The chapel had been whitewashed and well scoured and beautifully dressed in honor of the Bishop's visit, being all hung with holly, moss, and evergreens.

The sight was touching. All the women had met together and scoured first and then decked it with nature's wreaths. Poor little Alifair had brought her month-old baby, which weighed about five pounds, to be christened. She asked me to be godmother and I found she had named it after me. Her mother told me this and asked anxiously: "Miss Patience, you won't mind?"

"No, no," I said; "it is always a compliment." But when I found it was a combination of my name with my dear mother's I felt a little abashed.

The little thing was so very white and so very tiny that I felt almost afraid to handle her, poor little mite. The father, a boy of 18, found matrimony too serious and slipped away some months ago.

The church was very full and I think they all carried away real help in their daily lives from the Bishop's sermon. We did not get home until nearly dark, but it had been a most delightful day.

Goliah has been behaving very badly lately. Chloe came to me looking very portentous one day to tell me he had a gun hid in the kitchen in the yard and that when anything went against him or she told him to get more wood he would bring out the gun and threaten to shoot Patty and herself.

Goliah was out at the time and I asked Patty if she knew where the gun was. She said yes, and I told her to bring it to me, which she did, and I locked it up. When Goliah came home and missed the gun from its hiding-place he went on terribly, Chloe said, cursing and swearing to kill every one.

I was busy in the house and heard nothing of this, but finally he came to me before he went home and said Patty had taken his gun and please to make her give it to him. I answered:—

"I took the gun which has made you behave so badly and have locked it up."

"De my Brudder Bill gun, en' I want um."

"When your brother wants his gun he can come for it, but I cannot permit you to have a gun in this yard. Never again dare to bring one!"

He went off very sulkily. The next day Bill's wife came and asked for the gun, saying it was her husband's. I told her if I gave it to her it must be on condition that she never let Goliah have it again. Lizette was with her and I appealed to her to witness to Dorcas's promise never to let Goliah have the gun. The promise was made and I gave up the gun. Since then quiet has reigned in the yard.

I do not know how old Goliah is, but he is four and one-half feet high, and when he is good, like the little girl we have heard of all our lives, he is very, very good, and when he is bad he is certainly horrid. Now he is in a bad spell.

December 18.

A beautiful morning. Ran out to the gate with a letter. When I got a bundle of mail and opening it saw a telegram, I sank down on the ground in fright — and sitting there read the message. Sorrow for those I love, and it is too late for me to reach them in time for the funeral.

While I was debating what to do a boy came up with a note from Miss Penelope: "Terrific fire burning around Peaceville. Miss Pandora and Miss Ermine were nearly burned out. Miss E. worked like a trojan to save it. Better look out for your premises."

I ordered Bonaparte to take Adam, Nan, and a wagon with hoes and rakes and drive out to the village and give all the help he could, and protect my yard if possible. As soon as I had given these orders I drove out myself to take a telegram to the mail.

I found Miss Pandora and Miss Ermine looking as though they had been through a great illness. The fire threatened for two days and a night, and they had fought it all that time, but it was now put out. While I was in the village the wind changed and the fire appeared in another direction. I had Jim take Ruth and Marietta out and tie them in my yard, so that he could go and help put out the new fire. I took a young pine sapling and helped beat it out.

It is awful to hear the roar of fire through the pine woods and know how hopeless it would be if it once crossed the line and rolled around the rambling, elongated village. The negro men and women who happened to be near behaved very well and worked with a will, and I will certainly reward them generously. I stayed until the fire seemed finally out and by burning a space to meet the oncoming flames, I trust the danger is past. Those who are accustomed to wood fires, and there are men in the village who have had experience and directed the work, now think it is safe.

I sent a telegram to ask if I was needed, and if an answer comes I will go; if no answer comes I will know they do not need me. It would be difficult for me to go just now, as I am expecting some sportsmen as paying guests, and must be at home. I am as nervous as though I were going to execution.

CHEROKEE, December 22.

I was sitting at dinner to-day when I heard Goliah run up the front steps and ring the bell violently. Knowing from the sound that there was something the matter, instead of sending Patty, who is in a constant state of war with him, I went out myself. He was panting as from a long run and gasped out: —

"Pa baig yu for meet 'um to de front gate."

"When?"

"Rite now, soon ez yu kin git dey."

"Where is your father?"

"Him dey 'een de sheriff buggy gwine to de chain gang."

"How is that possible?"

"Dem had um to Mr. Haman to de co't dis mawnin'."

"Who had him there?"

"Dorcas en him ma, en pa dun condem'."

I waited for no more, entirely forgetting that I had not eaten my dinner of delicious shad. I put on my hat and flew out to the front gate. One feels very tenderly for those one has helped and poor Elihu has been a care and anxiety to me for years.

His family relations have always been difficult and complicated. His second wife and nearly all his children having died of a galloping consumption, he has now taken a third wife who has eight children of her own and is a termagant, so that though he works hard, and is honest and law abiding after a fashion, and very civil, even courteous, in manner, he is always in some trouble, generally debt. On my way to the gate I met Lizette, his daughter, crying as she ran on her way to call me. I asked what it meant, as I had been unable to get anything out of Goliah, and she explained to me that Elihu had been away working in a lumber camp when Goliah had threatened Chloe with the gun, and when he came home and heard of it he was very angry, and said Bill had no right to let Goliah have the gun; that he had lent it to Bill when he was getting well from typhoid fever and able to walk about and shoot, because Bill's gun had been taken for debt, he never having paid for it after the first instalment.

Yesterday Elihu went to Bill's house and asked for the gun. Bill was out and his wife refused to give up the gun, upon which Elihu scolded her, no doubt in strong language, until finally she went to her trunk and took it out piece by piece, trying her best to convince him that she did not have all the parts. She ended, however, by giving them all up. When her mother came in she reviled her for being so meek spirited as to take a scolding from Elihu and give up the gun and proceeded to curse and abuse Elihu.

Of course he was not found wanting in retorts and a neighbor had to come in and make the peace. Elihu thought nothing more of it until he received a summons this morning from the so-called Judge to appear in court. He went with no idea of anything serious, had no witnesses even, but found that Dorcas's mother had indicted him.

By this time we reached the avenue gate and I sent Lizette to run to tell Jim, who was ploughing in the field, to put Ruth in the buckboard quickly and bring her to the gate. Shortly after I reached the avenue gate a buggy drove up containing Mr. Stout, the deputy sheriff, and Elihu, looking too downcast, black, and forlorn for words.

Elihu is of a peculiarly rich shade of black, almost blue black. His own mother when he was a boy always spoke of him as "dat black nigger." Through all the trials and tribulations of his fifty years of life he has never been in danger of the chain gang before, for he has kept a good character for one of his hue, and now the certain prospect of the gang unless some miracle happened had crushed the spirit out of him. I scarcely would have known him. I walked out of the gate and said: —

"Why, Mr. Stout, what does this mean?"

"It means, Miss Patience, that I'm a-taking Elihu to the chain gang. I've got the warrant in my pocket."

"And on what ground?"

"For cursing, Miss Patience, and making a disturbance on the public highway."

"Was he not in his son's house?"

"Yes, Miss Patience, but the Judge says that is within fifty yards of the public road."

"Has it been measured, Mr. Stout?"

"No, ma'am, Miss Patience, 'tain't been measured, but the woman said it was only forty yards from the road, en the Judge said he knowed the place and that was right."

"What is the sentence?"

"Thirty days on the gang, Miss Patience, or a fine of $50."

"Mr. Stout," I said, "you turn right round with me and drive back to Mr. Haman with Elihu. That house is more than fifty yards from the highway."

This he said he dared not do.

By this time Jim had brought Ruth in the buckboard, and I got in and drove out of the gate.

"Mr. Stout," I said, "I thank you very much for having driven this way so that Elihu could see me, and I have a favor to ask of you. If you are afraid to go back with me, at least promise me you will wait at the turn of the road, until I come back. I will drive fast, you won't have long to wait, but you must do it," and before he could answer I had driven off.

It was a very cold evening and nearly dark. In my excitement I had put on no extra cloak, but I did not feel the cold. In marvellously short time the four miles were passed and I stood at Mr. Haman's gate. Goliah opened it and I told him to wait there until I came back, as I did not want him to hear the conversation.

On my second call Mr. Haman came out and was of course much surprised to see me. He was most polite, and eager to invite me in.

"Come in, come in, Miss Patience, I beg you," he said, "this wind is too piercing for you. I beg you will let me tie your horse."

"Thank you, Mr. Haman, but I cannot come in. I have only a moment's business with you. I want you to give me an order to Mr. Stout to release Elihu."

With a smile of greatest indulgence he replied, "You know, Miss Patience, I'd do anything to oblige you, but my duty, my duty, madam, is my first consideration, and even for you I must refuse to do anything contrary to my duty."

"Tell me," I said, "of what he is convicted and what is the sentence?"

"He is convicted of cursing and creating a disturbance

within fifty yards of the public highway, and the sentence is thirty days' work on the chain gang or a fine of fifty dollars."

"Mr. Haman, you must sign that release until the distance is measured. I know that it is more than fifty yards from the road to Bill's house, and until that distance is measured his committal to the gang is illegal. Elihu is a hard working, docile, respectable negro. If I wanted anything hard done to-night such as to send by land or water ten miles Elihu is the man I could call upon, knowing he would not refuse. If I had occasion to drive forty miles this night through the darkness, Elihu is a man I could trust to take me safely through the darkness and do it cheerfully. And you think I will see him put on the chain gang illegally? You don't know me, Mr. Haman."

He listened as if he did not hear, so determined was he not to yield and so accustomed to shake the law at people.

He said he would get the book and read me the section, but I said that was unnecessary, I knew the law; the point was whether this case was within the legal distance. Darkness was coming and I was making no headway. At last I said:—

"If I were to sign a note for fifty dollars would you give me the order for his release?"

"Oh, yes, Miss Patience, if you pay the money that'll be all right."

"Very well; bring me ink, pen, and paper and I will sign."

He went in and returned very quickly with pen, ink, and a check. I had not meant to sign a check, but a note; however, I signed it in ink and then asked for a pencil and on the back wrote, "Not to be presented until distance is measured." He seized the check with delight; when he turned it over and saw the writing on the back his face changed.

"Now," I said, "will you give me the order for Elihu's release?"

"No," he said, "I will not."

I was still sitting in the buckboard and I just leaned forward and took the check from his hand. He was so taken by surprise that he was silent for a second; then he said: —

"I'll go down and measure the distance for you, Miss Patience."

"When?"

"To-night, right now. I'll get my buggy en you kin go right on and I'll follow you."

I was truly thankful, for it was getting very late and I was so afraid that Mr. Stout would not wait. I drove rapidly toward the gate, which is approached by a causeway. When I got well on that, a thought struck me; though it would be a singular trial to me, to save time I would offer to take Mr. H. down in my buckboard. I saw a negro woman near and said to her: —

"Please run in and say to Mr. Haman that Mrs. Pennington will be glad to offer him a seat in her buckboard and he can come back with Mr. Stout."

She ran off briskly and in a few minutes returned and said, "Mr. Haman say never min'; say him ain't goin'."

Fortunately I had taught Ruth to back all over the yard before a harness was ever put on her, for I backed her the length of that causeway in no time and was back at the house. Mr. Haman came out looking considerably worried.

"Mr. Haman, you will not get rid of me to-night until you have signed that release."

"I can't do it! When I write a warrant it's writ, and everybody that knows me knows that."

At this juncture his wife appeared and said: "Miss Patience, he ain't well and it's too cold for him out here; please, ma'am, to come in."

I answered: "I cannot come in, Mrs. Haman. I simply want your husband to write an order to Mr. Stout to release

Elihu Green, whom he has sent to the chain gang for thirty days, until the distance from the road is measured."

"You're right, Miss Patience," his wife answered, and turning to her husband said, "Better do what Miss Patience wants you to do, an' come in out o' this cold wind."

Most reluctantly and heavily at last the words came: "You give that check to Mr. Stout en tell him to turn loose the nigger."

"And," I said, "you will send down early to-morrow to measure the distance?"

He was already disappearing in the door but assented, and again I started for home. Ruth by this time had got worked up and needed no whip; she knew there was something unusual in the air and she flew.

When I reached the turn by St. Cyprian's church, where Mr. Stout had promised to wait, it was so dark I could not see whether he was there or not until I came right up to him. There he was still in the buggy, and when I called, "Please come here, Mr. Stout," slowly he got out and came; I handed him the check.

He struck a match and examined it, then his whole face beamed and he said, "Then I kin turn Elihu loose?"

"Yes," I said, "and I thank you with all my heart for waiting; you have helped prevent a great injustice. Mr. Haman says he will send you down to-morrow to measure the land. If it is more than fifty yards, you will return that check to me; if less you will give it to him. Please come early."

Elihu was dazed with the sudden release as the handcuffs were taken off. After Mr. Stout left I gave Elihu a talk about the disgrace of cursing and making a disturbance, and I said : —

"Elihu, the distance will be measured and if it is less than fifty yards you will have to try and work that $50 out."

"Oh, yes, miss, de Lawd bless you, en I thank you too mutch, en I'll do all you want me to do."

And I made the best of my way home with Goliah behind. All the servants were wild with delight when they heard the result, and Chloe had a nice supper for me, but I was too tired to eat. The depths had been stirred within me and I could only go to the piano and play Rachmaninoff's grand prelude over and over until I was quieted.

Mr. Haman, the magistrate, is a man of foreign birth, and speaks broken English, a German, I suppose. He drifted here after the war and married the daughter of a very excellent old German who had bought a plantation and settled here at that period of change and unrest. Two years ago he was elected to his present office to the surprise of every one.

December 23.

The household was astir early this morning. As I felt it was a moment when I would like to have a gentleman with me, immediately after breakfast I drove over to my good neighbor Mr. F. and asked him if he would come with me and see the distance measured.

He said he would come with pleasure, and he got into my buckboard. Mr. F. with Jim's help measured and found the distance from Bill's house to the public road 250 feet, more than eighty yards. I was greatly relieved, for though I can generally trust my eye for distances I never had thought of this special space and had nothing to compare it with in my mind. It was simply an impression, I may say a conviction; and if it was wrong I would have to borrow $50, for I had not that much in the world that I could put my hand on at this moment.

Still I would rather do that than have Elihu punished and disgraced when he was really in the right.

Just as Mr. F. had finished measuring, Mr. Stout the

deputy sheriff drove up in a very fine buggy with another white man. I greeted them pleasantly and begged them to measure the distance at once, without saying that I had relieved my mind by doing so already. Mr. Stout assisted by Mr. Oliver took the measurements and pronounced it 250 feet. Then Mr. Stout handed me my check for $50.

Thank the good Father for His mercies.

Christmas Day.

Drove to church, where we had a pleasant service. It had been given out that the collection would be for Sewanee. There was great excitement after service when the word was passed around that it was $7! Our plate rarely holds more than half that amount. Every one was very happy over it.

Then went to take the few things I had gathered up for the St. Peter's people to K.'s to be sent to them.

I had to go to the extravagance of buying a comfort for the poor Lewis family; the weather has been unusually cold, and they are so destitute.

I have been quite alone to-day, but not at all lonely, for I have put up candy for the children on the place and little packages for the old people. To-morrow I am to have the joy of a visit from my two nephews, one of whom has been living in New York and has not been here for a long time.

December 26.

E. and A. came about 1 o'clock with guns and dog — perfectly charming both of them — both full of zeal to shoot. I sent Jake up to get another boat and engaged him to come at 5 A.M. to-morrow to take them out. He is to get Aaron, who is a good paddler.

December 27.

Very early this morning Jake came and said he had failed to get Aaron. Jim came into the house, made the fires, and waked the boys; then went out to the "street" to get a paddler. First he went to Frankie's house; he was in bed and refused absolutely to come, saying he was too comfortable. Then he went to Gibbie's house; he talked a great deal, but finally said he would come.

Meantime E. got off in the light canoe with Jake. A. waited until Gibbie rode up on his bicycle to say that it was a very dangerous business, etc., and finally that he could not come, so Jim had to take A. out, though he is no paddler.

As soon as I heard what had happened I wrote a paper to Frankie and one to Gibbie. "You will leave this place at once," and sent it to them. Only yesterday when they came up for the annual powwow they made a solemn promise to do all that I called on them for. I made them promise this, because for some time there has been a growing disposition not to do what I want done, and if I let it go on and pass over anything like this I will lose all control of the place.

I am sitting out in the sun and have thawed out while writing this. Oh, the goodness and mercy of God! A sense of it pervades my being to-day. Though I have had my small trials already this morning, they seem as nothing when I think of all His patience and long-suffering and loving-kindness to me.

CHEROKEE, December 28.

E. and A. seem perfectly happy to be here, and their visit is an unbroken pleasure to me. They have not got very much game, but just to paddle round the creeks or to walk over the woods gun in hand seems to revive all the happiness of their childhood.

To-day we went to Casa Bianca for the day, and went prepared for them to spend the night if they found the ducks plentiful enough to make it worth their while. When they went down into the fields, Nat paddling one and Jim the other,

I started back home alone. I got here after a perfect drive and glorious sunset.

To my great pleasure I found C. here, and then began to hope the boys would not spend the night at Casa Bianca. So I was delighted when at 9:30 the dogs announced their arrival. They had shot a good many ducks, but not having a dog to fetch had not got one.

We had a delightful evening. While I was gone to-day a man had brought a bushel of oysters fresh from the sea, so I could give them a nice oyster supper.

Rice-fields from the high lands.

Gibbie came to-day and among other things said he never could paddle. "Yu knows yuself, Miss Pashuns, I neber could paddle."

"On the contrary, Gibbie, I know you to be an expert paddler," I replied. "So much for that excuse.

"Three of your children were born in the house I had repaired for you. Then the best house on the place became vacant and you asked me to let you move into that because it was nearer your brother's. I let you do it and charged no more rent. You were ill. I paid your doctor's bill, which money you have never returned to me. I sent you milk daily until you were quite well, and during your mother's illness of three months I sent her a pint of fresh milk night and morning until she died.

"Your wife this winter burned down the house you occupied. There is no use for you to shake your head and say no. What else was it when she sent your eldest boy, 5 years old, into the loft, which was packed with fodder and corn and hay, to get peas, with a burning lightwood torch in his hand?

Instead of telling you that you must leave the place, as I could not furnish another house to one so criminally careless, when you begged me to have another house repaired for you to move into, I did so. The burning of the house was a complete loss of $300 to me.

"I let you off the rent of the house for the last month and waited on you for the payment on the other house long after it was due. All this time you told me you had no money and I waited, but you told other people on the place that you had $30 put away from your last three months' work.

"It would take a whole day to remind you of all the kindnesses I have done to you and all the meannesses you have been guilty of to me. And now my father's two oldest grandsons come to spend a few days at the old home shooting ducks, and I send for you to paddle one of them in a boat, not as a favor, mind you, but to be well paid for it, and you ride up on your bicycle to say you cannot do it.

"There is no use to say a word, Gibbie, I will not hear it. The time has come for us to part company. You must go."

He turned and walked down the step with a sullen look. I will miss the $2 a month which the two houses brought in, but it is necessary to do a thing like this once in a while.

CHEROKEE, December 30.

Mrs. L. has become greatly interested in the poor white people out at St. Peter's-in-the-Woods and she sent me word that if convenient to me she would come up to-day in her motor and get me to go out there with her to distribute some things which she had collected for them. I was so delighted at her interest that I said it would be perfectly convenient, and though in the back of my mind was the picture of the dining-room chimney place all torn to pieces, I asked the party to take lunch with me.

So early this morning I sent for Bonaparte and told him he

must make some mortar and repair the fireplace and put back the mantelpiece and please to have it done by 12 o'clock so that Patty Ann could clean up the mess and make the fire and be ready for lunch at 1. All of which was done, and by the time the sound of the auto was heard everything was ready but myself.

I had been obliged to contribute greatly to the result and had not time to change my working outfit before they came. That did not matter, however. They brought a huge hamper and basket full of all sorts of nice things. The dear little girl had brought lots of her dresses and above all toys! Such beautiful things, Teddy bears and billikins and dolls and animals and clowns. They brought also groceries.

We had our lunch and then I joined them and we went the nine miles in no time. The visit to Louise Moore was most successful. She and her house and children were clean and sweet. That term could not be applied to the biggest boy, about six, however, as he had been skinning four possums, which were extended on sticks in the little porch.

Then we went on to poor old Mrs. Sullivan and her Dickens-like daughter. She was overjoyed at the groceries and nice things. Her great poverty was very apparent in her surroundings, above all the flimsy garments she wore, but all was clean. The next visit, two miles beyond, was also satisfactory, but alas, the last visit was a shock. Mother, daughter, and granddaughter were too untidy for words. I could not help wishing we had not gone there, it was so disappointing.

Certainly nothing could show more their need of help and industrial training. I had only seen them as a rule at church and had no idea this special family was so untidy. I had been to the home two or three times, but I suppose that was not on Saturday afternoon when everything, including ablutions, had lasted over since the Saturday evening before.

This last visit rather dampened our spirits, though the little Frenchwoman, who had carried a large box of stick candy which she distributed as we went, found something pleasant to say even about that.

When we got back to Cherokee Chloe had a cup of tea ready and the party returned to Gregory. I felt anxious, it being late and cold. They left a large basket of things for me to keep for further distribution. I wish so I could get at the poor Lewis family with some of them.

Miss Chevy, who was visiting Mrs. Sullivan, answered when I asked about the Lewises in a high and righteous voice: —

"Yes, Miss Pashuns, they've gone away bag and baggage an' I tell you truly it's a good riddance, Mis' Lewis she acted that ridiklous with them children.

"A man come there one day in a wagon from de up country lookin' for han's to pick cotton, an' he asked me if them Lewises could pick cotton, an' I spoke up an' said, 'Yes, sir, they kin pick cotton every one o' them, 'en he jes' drove right to the house an' asked them to go with him en he carried them all off, father and mother and three children, en I'll tell you, Miss Pashuns, it's me that's thankful.

"You see I didn't tell no lie; he didn't ask me if they would pick cotton, but he ask me if they cud, an' I up an' says they cud, but I didn't say they's that shiftless that they won't do it."

"You see I didn't tell no lie."

In the mail which I found when I got home I had a letter from a friend referring to an adventure which I had four years ago that I do not think I ever wrote down, so now I am going to do so, for I forget things so entirely. My friend had come from New York to make me a visit of a week. At the end of that time, wishing to be with her as long as possible, I drove her to Gregory in the buckboard to take the train. The train left at 4:30, which in the latter part of December is very nearly dark.

I had taken Jonadab behind the buckboard. When I started on the fourteen mile drive home, I felt dismayed, for I knew it would be dark soon. I crossed the ferry with the last light of the dusk and drove on into blackness.

I had only gone a little way, however, when we drove into a forest fire. Both sides of the road were aflame and Ruth at first was frightened, but finding it did not come into the broad, white, sandy road, she soon enjoyed the illumination as I did.

For about three miles we passed through this brilliant region, and then I saw we were coming to the end of it and would soon be in the darkness again, so I told Dab to get out and pick up a good piece of lightwood for a torch and light it, which he did, succeeding in a very short time in getting a long, fat piece full of turpentine, and just before we left the fields of light he lit it and held it behind so that it gave a very satisfactory path of light just ahead of the horse.

All went well until we came to a turn in the road where we had two bridges to cross and I feared, as they were narrow and without railing, that I might not see well enough, so I told Dab to get out and carry the torch in front until we crossed the bridges. This he did, walking quite rapidly, so that just after we crossed the last bridge the torch blew out; the rapid motion somehow being in front, made too much draught. Dab was much concerned, but I said: —

"It does not matter now, Dab, we are only three miles from home and I know every foot of the road; get up behind and we will soon be at home."

He got up behind the buggy and we went on several hundred yards, when there was a terrific report, and great flames of fire, blue, green, and red, passed over our heads. Ruth dashed, throwing me out on my head, upsetting the buggy, broke loose and disappeared in the darkness. I must have been stunned, for I got up quite confused, found no horse in the shafts, and just walked ahead, forgetting all about Dab.

As I walked on I heard the noise of the horse in the woods to the left of the road. I went in a little way and called to her. Fortunately she has always come to my call, and did not fail to do so now. I caught her and led her back to the buggy. I found both traces broken and felt hopeless.

By this time Dab, who must have been stunned too, came forward to help. I gathered all the strings that the resources of feminine apparel furnish, and tied up the traces, then without getting in, told Dab to lead Ruth off, which he did, but the buckboard did not move. I had no knife to cut holes in the leather, so no string could hold. Still making the effort to secure the two pieces together I said: —

"Dab, what did happen? I never have seen or heard such a thing before. Do you now what it was?"

Dab, stuttering fearfully, said: " 'Tis — is — is — is de fiah cracker, ma — a — a — m!"

"What?" said I.

"Yes, ma — a — m, I — I — I buy six roman candles to town en I had dem een my bussom en me jacket button ober dem, en w'en the torch gone out I ben' down, en bin a blow um fu make um blaze, en a spaa'k fly een me bussom en set off de roman candle, en den dem blow we up."

There in the darkness three miles from home, with no hope of mending the harness, I laughed until I sank on the sandy road. I could not stop laughing, to Dab's great amazement. Why his nose was not blown off I can't imagine; it had been such a near thing that he was much nearer tears than laughter, and he had expected certainly a scolding from me, and now this totally unexpected and unnatural laughter awed him still more.

When I resumed my efforts I saw far down the road a light drawing slowly near. When it got within hailing distance I called several times before I got an answer. I said: "Please come here, whoever you are. I am in trouble and I want your help."

They seemed reluctant and came slowly. When they got near enough and the light fell on me, one man said: "Why, my Lawd, 'tis you, Miss Pashuns?"

"Yes," I said. "Who are you? I don't seem to know you in the dark."

"No, ma'am, you don' know me, but I knows you well. I'se Rastimus en dis is my fren' Joshuay."

"Well, Rastimus, I want you and Joshua to fix my harness for me. I've had an accident and I can't manage it at all myself. Have you a knife to cut a hole in the trace, because we can do nothing without that?"

"Yes, my missus; I got a very shaa'p knife, en don't you worry, me en Joshuay'll hab um fix korrek, fo' yu knows it."

And true enough, though their motions were very unsteady and the air was redolent of firewater, in very quick time the harness was tied up in an ingenious and substantial way. Then I asked for the loan of the lantern. This they hesitated about, but when I gave my word that it should be sent to the store the next day with a little note of thanks and an enclosure for each, they consented, and I went on my way with songs of praise and thanksgiving in my heart for the many and varied dangers I had escaped. The next day the lantern was duly returned, with a quarter apiece for my knights-errant.

December 31.

Spent this last day of the old year writing letters of thanks and affection, and after dark I made up a bright fire, Chloe and Patty Ann having gone away on their Sunday outing, and sat in the firelight without lighting the lamps and reviewed the mercies and blessings of the past year. God forgive me for my mistakes and sins therein, my blindnesses and lost opportunities.

I keep wondering if it is His will that I should give up this life. I do not want to be headstrong about it. I have so

loved the freedom and simplicity of the life, in spite of its trials, and isolation. The living close to Nature — the trees, the birds, the clouds, and all the simple loving dumb things.

But it almost seems as though I was meant to give it up. The rice-planting, which for years gave me the exhilaration of making a good income myself, is a thing of the past now — the banks and trunks have been washed away, and there is no money to replace them. The experiment of planting cotton has not been a success with me. The cotton grew luxuriantly and bore well, but others gathered it, and I got but little. I cannot sit idle in the midst of all this fertile soil. But I must wait, and watch, and listen, in silence, for the still, small voice, which comes after the storm and the earthquake, and brings the message from above.

THE JOHN HARVARD LIBRARY

The intent of Waldron Phoenix Belknap, Jr., as expressed in an early will, was for Harvard College to use the income from a permanent trust fund he set up, for "editing and publishing rare, inaccessible, or hitherto unpublished source material of interest in connection with the history, literature, art (including minor and useful art), commerce, customs, and manners or way of life of the Colonial and Federal Periods of the United States . . . In all cases the emphasis shall be on the presentation of the basic material." A later testament broadened this statement, but Mr. Belknap's interests remained constant until his death.

In linking the name of the first benefactor of Harvard College with the purpose of this later, generous-minded believer in American culture the John Harvard Library seeks to emphasize the importance of Mr. Belknap's purpose. The John Harvard Library of the Belknap Press of Harvard University Press exists to make books and documents about the American past more readily available to scholars and the general reader.

Bei Fragen zur Produktsicherheit wenden Sie sich bitte an:
If you have any questions regarding product safety,
please contact:

Walter de Gruyter GmbH
Genthiner Straße 13
10785 Berlin
productsafety@degruyterbrill.com